NATURAL SELECTION AND GENETIC DRIFT

GENETICS - RESEARCH AND ISSUES

GENETICS - RESEARCH AND ISSUES

NATURAL SELECTION AND GENETIC DRIFT

JOSHUA RICHARDSON
EDITOR

nova publishers
New York

NOTICE TO THE READER

Library of Congress Cataloging-in-Publication Data

Names: Richardson, Joshua, editor.
Title: Natural selection and genetic drift / editor, Joshua Richardson.
Description: Hauppauge, New York : Nova Science Publishers, Inc., 2016. |
  Series: Genetics--research and issues | Includes index.
Identifiers: LCCN 2015044952 (print) | LCCN 2016000014 (ebook) | ISBN
  9781634843317 (hardcover) | ISBN 9781634843324 ()
Subjects: LCSH: Variation (Biology) | Natural selection. | Evolution (Biology)
Classification: LCC QH401 .N38 2016 (print) | LCC QH401 (ebook) | DDC
  576.8/2--dc23
LC record available at http://lccn.loc.gov/2015044952

Published by Nova Science Publishers, Inc. † New York

CONTENTS

Preface — **vii**

Chapter 1 — Processes in Organic and Cultural Evolution — **1**
Börje Ekstig

Chapter 2 — Natural Selection and Diabetes Mellitus — **61**
Svetlana Shtandel

Chapter 3 — Genetic Drift among Native People from South American Gran Chaco Region Affects Interleukin 1 Receptor Antagonist Variation — **101**
*Cecilia Inés Catanesi
and Laura Angela Glesmann*

Index — **119**

PREFACE

Natural selection is the process which, being the most important factor of evolution, promotes rising of adaptability and prevents destructive consequences of all other processes. The concept of natural selection is a discordant problem of evolutionary human genetics. Despite popularity of a hypothesis of "neutral evolution," the majority of scientists consider that selection has played main role in evolution of species and has generated all bio-logical diversity of human populations. This book presents research on natural selection and genetic drift. The author of the first chapter provides an all-embracing macroevolutionary perspective on the processes of the evolution of life and culture on earth. The author investigates a complementary form of natural selection that diverges from the traditional form in that it is acting independently of the external environment. The next chapter discusses natural selection and diabetes mellitus. The last chapter examines how the genetic drift among native people from South American the Gran Chaco region affects interleukin 1 receptor antagonist variation.

Chapter 1 – In the present work, the author gives an all-embracing macroevolutionary perspective on processes of the evolution of life and culture on earth. First, the author investigates a complementary form of natural selection that diverges from the traditional form in that it is acting independently of the external environment. This form of natural selection is found as a result of a mathematical analysis of the conditions for population growth. The author extends the author's investigation as well to other evolutionary processes than the organic, such as the evolution of human language and the evolution of science, thereby suggesting other possible forms of underlying explanatory processes. The author examines the concept of complexity and show that it implies new insights into the ways natural selection has been acting in forming the evolutionary and developmental processes. Especially, complexity is found to be growing in an accelerating mode, a process that is explained as the combined result of natural selection and a self-reinforcing feedback process. The use of the concept of complexity opens the possibility of construction of a new form of a Tree of Life, which, in contrast to traditional forms, combines complexity and time. Such an illustration of the evolutionary process explains the observation that most species live without great changes over vast periods of time. For species at the highest level of complexity there is no competition from species at still higher levels and these species can therefore, if conditions are beneficent, form new species at still higher levels. The process explains the emergence of new species and the general trend of evolution towards cumulatively higher complexity levels. The cumulative addition of species with successively higher complexity implies that the latest appearing species is the one of the highest level of complexity. At present, this species is the human species.

Chapter 2 – Advances in modern medicine enable a change in the tension of intragroup selection in human populations. Thus, implementation of insulin for type 1 diabetes mellitus (DM) treatment considerably lowered the selection tension for this symptom and converted it from the sub-lethal to the one with a lowered adaptability. Increasing variety of type 1 DM and type 2 DM is being observed recently in different populations. Moreover, recently the heterogeneity of type 1 DM and type 2 DM has also been observed. The investigation was aimed to study the influence of the selection on the evolution of DM clinical forms. Global implementation of insulin therapy into type 1 DM treatment caused the prevalent increase of this disease. Currently, there is a positive selection trend for type 2 DM, which is the original cause for the prevalence increase within the population, and the negative one for type 1 DM determines its prevalence within the population approximately on the same level. Intra-population change of gene frequencies, susceptible for type 1 and 2 DM, predetermined the development of such DM clinical forms as LADA (latent autoimmune diabetes in adults). It also resulted in the increasing number of patients with an absolute insulin deficiency of type 2 DM, which is a more complicated form of this DM type. Polymorphisms association, changing the immune response and forming the susceptibility to type 1 (C1858T of the gene *PTPN22*, A49G of the gene *CTLA4*) and type 2 (E23K of the gene *KNJ11* participating in insulin insufficiency formation) with different DM forms illustrates the result of decreasing the selection tension against type 1 DM after insulin therapy implementation into the public health services practice.

Chapter 3 – Genetic variation is generally responsible for ethnic differences in certain diseases, including inflammatory processes. The antagonist of cytokine IL-1, IL-1Ra, has been widely studied among Caucasian and African populations for genetic polymorphisms, and interethnic differences have been documented. However, the variation and genotype distribution of polymorphisms from these genes among South American Amerindians are thus far unknown. The author's present the results for a VNTR located in the IL-1Ra second intron, in a sample of 169 individuals belonging to 5 Native American populations from Argentina and Paraguay, identified as native according to their self designation, and their geographic location. The author's also compare this data with the results obtained from a sample of non-native Argentinian people. Among the five known alleles of the VNTR, the author's found only two (alleles 1 and 2) in the native populations from Gran Chaco, and heterozygosity was 19%. The allele 2 which is considered proinflammatory (IL-1Ra * 2) has been found in homozygosity at a considerable frequency among native individuals. However, the association of this allele with inflammatory disease previously demonstrated for other populations of the world, might not be acting in the same way for native people, probably due to local adaptation. This would indicate that the allele 2 will probably not have a negative influence on individuals of native origin who have homozygous genotype 2-2. On the contrary, few records on inflammatory disease are available for the native people. It seems that the increment on allele 2 is not related to any adaptive process but to genetic drift, that changes randomly the allele frequencies of different genetic regions along the genome. The effect of genetic drift has already been demonstrated with genetic markers located in autosomes, X and Y chromosomes. These results indicate that the author's must be very cautious when studying populations that passed a process of genetic drift, which can become a confounding factor in epidemiological studies. This information will contribute to a future understanding of the association of this polymorphism with disease, and its incidence in different ethnic groups.

In: Natural Selection and Genetic Drift   ISBN: 978-1-63484-331-7
Editor: Joshua Richardson   © 2016 Nova Science Publishers, Inc.

Chapter 1

PROCESSES IN ORGANIC AND CULTURAL EVOLUTION

Börje Ekstig[*]

Department of Education, Uppsala University, Uppsala, Sweden

ABSTRACT

In the present work, I give an all-embracing macroevolutionary perspective on processes of the evolution of life and culture on earth. First, I investigate a complementary form of natural selection that diverges from the traditional form in that it is acting independently of the external environment. This form of natural selection is found as a result of a mathematical analysis of the conditions for population growth. I extend my investigation as well to other evolutionary processes than the organic, such as the evolution of human language and the evolution of science, thereby suggesting other possible forms of underlying explanatory processes.

I examine the concept of complexity and show that it implies new insights into the ways natural selection has been acting in forming the evolutionary and developmental processes. Especially, complexity is found to be growing in an accelerating mode, a process that is explained as the combined result of natural selection and a self-reinforcing feedback process.

---

[*] Corresponding author: Börje Ekstig. Uppsala University, Department of Education, Uppsala, Sweden. E-mail: borje.ekstig@gmail.com.

The use of the concept of complexity opens the possibility of construction of a new form of a Tree of Life, which, in contrast to traditional forms, combines complexity and time. Such an illustration of the evolutionary process explains the observation that most species live without great changes over vast periods of time. For species at the highest level of complexity there is no competition from species at still higher levels and these species can therefore, if conditions are beneficent, form new species at still higher levels. The process explains the emergence of new species and the general trend of evolution towards cumulatively higher complexity levels.

The cumulative addition of species with successively higher complexity implies that the latest appearing species is the one of the highest level of complexity. At present, this species is the human species.

Keywords: organic evolution, cultural evolution, natural selection, complexity, self-reinforcing feedback, the tree of life

INTRODUCTION

Individual Development and Evolution

Life is fantastic. To me it has always been irresistible to ponder upon the seemingly highly improbable course of events that has resulted in the appearance of mankind on earth and the likewise unfathomably long time during which this very process has incessantly been advancing. Equally enigmatic to me seems the marvellous developmental process of an animal or a human being, merely arising from the molecular genetic code, inherited from the long and unbroken chain of individual developmental courses ever since the dawn of life. In addition, it is to me the highest intellectual challenge to try to understand the intricate coupling between these two processes. Indeed, this is part what this work is about.

All the way during the course of evolution, natural selection is by the majority of researchers considered as the key mechanism. In this selection process, competition between organisms has been a decisive factor. Highly chaotic external factors have formed the underlying conditions for this contest in which natural selection has appointed the winner. The competition may have been about mates and food, as well as about capability to handle social life and the impact of diseases, predators and climate. Thus natural selection gives rise to a population's adaptation to its environment, thereby producing the enduring though irregular evolutionary changes. Due to the irregular nature of these external conditions it has been difficult, as testified by the history of the science of evolution, to find an all-embracing pattern and direction of evolution at large.

It is important to notice that the genetic material, the DNA, doesn't determine the evolutionary process. Instead, the DNA-molecule gives the plan for the individual's growth and its realization in the building up of an embryonic and juvenile individual creature. In the present work, I'm not entering the field of molecular biology – rather I discuss the macroscopic aspects of development and its evolutionary outcome.

A common notion of evolution seems to be to regard it as an extended sequence of adult organisms sustaining successive transformations due to natural selection. However, such a notion encompasses limited truth only. A central tenet forming the basis of the present thesis is that *natural selection primarily is acting on the developmental process* of individual growth and that *the process of evolution above all proceeds as a result of insensibly small and continuous modifications in the development programs* of living organisms, modifications having during the long extent of time given rise to the vast diversity of life on our planet.

This view is by no means new. It is expressed, for instance by Richard Dawkins, in saying that "evolutionary change is change in genetically controlled processes of embryonic development, not literal change from adult form to adult form (Dawkins 2004 B p. 200), and furthermore "that we have to understand development before we can speculate constructively about evolution" (ibid. p. 201). These aspects of development indicate the existence of some kind of relationship between development and evolution; a notion that has caused a long and engaged discussion ever since Darwin's discoveries. In the current work, I present my own discovery of a concrete manifestation of this relationship and my interpretation of it as well as some far-reaching outcomes of the discovery.

The Problem of Evolutionary Change

In a tradition that harks back to antiquity, living organisms have been seen as arranged in a hierarchical order, thus placing the most simple at the lowest level and the more advanced higher up. This principle was developed without any time dimension, because the world with all its organisms in their present shape was thought to have been created at one and the same occasion. During the end of the eighteenth century, observations of successive temporal changes of organisms became impossible to deny, a notion that got its fulfillment by Charles Darwin in his explication of this successive change, an explication that, after unprecedented turmoil, finally became accepted by the scientific community. As we know, Darwin called his principle natural selection and the process of temporal change became called evolution.

Then however, the idea of hierarchical order came to be considered as obsolete and became abandoned. This is by and large the situation of today's scientific conception of the evolutionary process, even if there are attempts to revitalize the notion of hierarchical order. Thus Daniel McShea (2001) has argued that an increase of the complexity of organisms in evolution can be measured on a hierarchy scale. In the present work, I intend to show the high explanatory possibilities of arranging organisms in a progressively rising complexity scale.

But the situation isn't that easy because most organisms show only marginal changes over long periods of time. How then, as one may ask, to explain this observation by means of natural selection, supposedly being acting on behalf of the great variations in the external conditions that doubtlessly have occurred? And what's more, how to explain the fact that the oldest organisms of today, that is to say those having been exposed to natural selection for the longest time, still retain their archaic characteristics? Actually, in the beginning of the nineteenth century, Lamarck pondered over this enigmatic observation already when the very idea of evolution was quite new and even before Darwin discovered the principle of natural selection. He asked how we still can see the complete hierarchy of living organisms today, if the active power of nature compels life to mount steadily up the chain of being. Why haven't all living things raised themselves to the same level as man? In his broad exposition of the history of evolution, Peter Bowler (1989 p. 85) has called attention to Lamarck's thought-provoking conundrum. Although much of Lamarck's views now are obsolete, I find his questions still challenging and my intention with the present thesis is to suggest an answer.

The reason for the scientific denial of a hierarchical order of organisms is obviously a lack of means by which to identify and measure such evolutionary levels. Such identification requires a measure of evolutionary progress, a measure that has turned out to be difficult to find.

However, there is a vast literature aiming at a solution of this deficiency. The most frequently suggested candidate for a concept to be used for the assessment of evolutionary progress is complexity. Before a discussion of this concept, I suggest that we prepare ourselves by making acquaintance with a couple of research fields associated to the evolutionary process.

Aim

The aim of the present work is to examine natural selection and its impact on the macroevolutionary process of life and human culture. Special attention is given to my discovery that natural selection in certain situations is acting independently of the external environment. The aim is furthermore to examine the concept of complexity, a concept, as is my intention to show, having great explicatory power for our understanding of the evolutionary process at large. Especially, I analyse the impact of a feedback process in complexity, a mechanism that I suggest as a complement to natural selection. I finally examine to what extension the discussed processes of biological evolution are applicable to human cultural evolution.

Evolutionary Developmental Biology

As I already have emphasized, my main view is to see evolution as the result of changes in the developmental course of individual creatures – by no means a new view. It is given attention in a special theory called Evolutionary developmental biology (*Evo-devo*). In his introduction to Evo-devo, Wallace Arthur (2011 p. 28) states that development is both a source for, and a result of, evolutionary change. Likewise, in his emphasis on the genetic background, Sean Carroll accentuates that the evolution of form occurs through changes in development (Carroll 2005 p. 4) and that molecular changes contribute substantially to organismal adaptation (Carroll 2008). In his summary of Evo-devo, Gerd Müller (2007) states that this field explores the relationships between the processes of individual development and phenotypic change during evolution. Müller discusses how developmental systems have evolved and probes the consequences of these changes for organic evolution. He contends that Evo-devo aims at explaining how development itself evolves and how the developmental processes are formed by the interplay between genetic, epigenetic and environmental factors. Unfortunately, the embryonic development of ancestral animals is not available for observations – they are not fossilized – and therefore conclusions regarding evolutionary trends of developmental changes are difficult. It seems that this difficulty to some degree is surmounted by the study of invertebrates. Thus Tills et al. (2011) have performed studies of snails taking advantage of the fact that snail embryos have benefits over those of vertebrates as they allow the study of the continuous developmental process.

Some authors point out that modifications in early embryos are very uncommon because of their supposedly profound effects on subsequent developmental events (Raff and Wray 1989). This view is even more clearly expressed by Brian Hall (2002 p. 13), suggesting that early development is immune to change or, if altered, would so drastically deflect subsequent development that early changes would be lethal or actively selected against. Yet, there is a type of changes that are not impeded by this obstacle. Thus, Kalinka and Tomancak (2012) contend that many changes in early development result in increase in the tempo of embryogenesis and a shift from late to early-patterning processes; a view studied in the field of heterochrony. These remarks are of high significance for the continued analysis in this work.

Heterochrony

As emphasized in the field of Evo-devo, modifications of the developmental programs of growing creatures provide an important entrance to the understanding of evolution. A special kind of such modifications is temporal displacements of the formation of developmental traits, a phenomenon called *heterochrony*. Central contributions to this field of research are given by Stephen Jay Gould's (1977) classical *Ontogeny and Phylogeny,* by McKinney and McNamara's (1991) *Heterochrony* and by Rudolf Raff's (1996) *The Shape of Life.*

In his summary of heterochrony, Kenneth McNamara (2002, 2012) defines this process as a change of timing or rate of development relative to the ancestor. He states that heterochrony takes the form of both increased and decreased degrees of development. He also recognizes that genes regulating embryonic or larval development play a major role in evolution. These genes are the targets of mutations causing heterochronic change in phylogeny. A link between evolution and development is formed by the inherent directionality in evolutionary trends also occurring in organisms' development. McNamara concludes that also human features are moulded by the all-pervasive influence of heterochrony. Our overall large body size, relatively large legs, structure of our pelvis, enlarged cranium and brain, and even our big feet can be seen as resulting from heterochronic processes. As Raff (1996 p. 267) points out, studies of hererochronies have predominance for late development, but he stresses that hererochronies in early development have been found to be as prevalent as in late development.

I would here like to remark that all heterochronic changes are more or less tacitly assumed to be formed by natural selection.

In the present work, I restrict the discussion to two simple forms of heterochronies – condensation and terminal addition. *Condensation* means the successive shortening of the interval of time needed for the shaping of each specific trait in the developmental process of individual creatures, a process causing displacement to earlier emergence of subsequent traits, a process called *acceleration*. Thus acceleration is caused by condensation and I therefore focus the discussion on condensation.

It should be emphasized that condensation, like other heterochronic processes, is a process expressed both in ontogeny and phylogeny, thus forming an issue for Evo-devo. One representative of this field of research maintains that when comparisons are made between very different levels of complexity, the emerging pattern is broadly recapitulatory, although only in a very imprecise way, in the sense of recapitulating levels of complexity rather than precise morphological details (Arthur 2002). It is interesting to note that Arthur includes complexity in his discussion, a concept that is central in the present thesis as well and will be analysed in forthcoming sections.

Terminal addition means the emergence of new traits with new functions at the end of the maturation period. One may think that terminal addition cannot be included in a strict definition of heterochrony since it means no change of timing or rate of development of existing early organs. Nonetheless, for instance McKinney and McNamara (1991) include it in their abundant terminology. They argue that much of the overall pattern of evolution can be explained by terminal addition and that one particularly important arrow of evolution, that of increasing complexity, seems to owe much of its existence to this process (ibid. p. 381).

Terminal addition occurs near the stage of maturity and is explained by the traditional interpretation of natural selection. An addition of a new trait is of course possible only on behalf of its reproductive advantage in the prevalent environment. Otherwise it wouldn't be inserted.

Terminal addition causes prolonged generation time owing to the insertion of novel traits at the end of the juvenile period. Such novel traits can lead to the emergence of new species and in this way terminal addition explains the diversity of life (ibid. p. 381).

It must in this context be mentioned that after the stage of maturation great changes are rare because after reproduction, the fate of the adult animal has no bearing on the genes of its offspring. Of course, this rule is not absolute in the case of the adult's caring of its offspring. Furthermore, it isn't abrupt for iteroparious species.

The two processes of condensation and terminal addition result in opposite evolutionary trends; condensation results in shortening of the time from conception to adulthood, i.e., a shortening of generation time; terminal addition on the other hand prolongs maturity time, i.e., it increases generation time. A common notion, as Stephem Jay Gould (2002 p. 367) mentions, seems to have been that new stages cannot be added indefinitely to the end of previous ontogenies, lest growth of adulthood take untold years to reach completion. Some process must therefore, as it is thought, produce general speeding-up of development, leaving time at the end for novel features. Such a view, I think, implies evolution to embrace an element of intentionality, a view that must be resolutely rejected. An important goal of the present work is to explain the speeding-up of development in a scientific way.

As to heterochrony in general, it maybe isn't that all-pervasive as McKinney and McNamara contend. Thus Rice (1997) proposes that although heterochrony has become a central organizing concept relating development and evolution to each other, it can only accurately describe a small subset of the possible ways through which ontogeny can change. The somewhat bewildering nomenclature of heterochrony is meaningful only when there is a uniform change in the rate or timing of the ontogenetic process, with no change in the internal structure. In the present thesis, such a uniform temporal change without change in the internal structure forms a basic principle.

As mentioned above, there is a generally expressed contention that changes of traits in early development would so drastically deflect subsequent development that such changes might be lethal. But a change restricted merely to an increase in the tempo of the shaping of a trait, without change of internal structure, wouldn't change its function and therefore wouldn't be prohibited. This is what characterizes the process of condensation. But there is another important consequence of this conclusion. If there is no change of function, natural selection has no alternatives among which to choose. How then to explain condensation? It seems that we have to examine the very process of natural selection somewhat closer.

ENVIRONMENT-INDEPENDENT NATURAL SELECTION

Population Biology

In his study of population biology, Stephen Stearns (1992) points out that a shortening of generation time is beneficial for the population growth because it increases its chances of surviving the often-risky juvenile period. A higher number of individual creatures will reach maturation and get the possibility to contribute to reproduction. In addition, a shorter generation time implies more frequent occasions of reproduction over time.

Therefore, these double advantages generate a selection pressure on all developmental traits to develop as rapidly as possible with retained functionality. Such a selection may give rise to a shortening of developmental traits once they have been established. We recognize this shortening as condensation.

There is simultaneously also a benefit of the opposite process – the prolongation of generation time. A delayed maturity, as Stearns points out, may imply greater size and a higher quality of the offspring, which leads to higher initial fecundity. This process, I maintain, is only occurring at the end of maturity time as terminal addition, since great changes of early developmental traits often are deleterious. Such terminal additions are explained by the traditional interpretation of natural selection.

Thus, there are two counteracting processes influencing generation length; one causing its shortening, associated to condensation of developmental traits, the other causing its prolongation, associated to terminal addition. An important contention of the present analysis is that these two processes are acting independently of each other. Hence, even if a species prolongs its generation time by means of terminal addition, its developmental traits may nonetheless be subject to condensation, implying that the actual change of generation time is the total result of both processes. In the majority of cases, the prolongation dominates but an assessment of the contribution of each of the processes seems difficult.

As we may conclude so far, there is a selection pressure on all developmental traits to develop as rapid as possible. An important question is now if the traditional form of natural selection is sufficient for explaining such a shortening because, as I already have pointed out, if there is no change of function, natural selection gets no alternatives among which to choose. One may thus conclude that condensation is invisible to natural selection, at least in its traditional interpretation. There is need for a more detailed analysis.

A Mathematical Analysis

The question is if the traditional form of natural selection is sufficient to explain condensation of developmental traits. I suggest an investigation of this question by means of a simple mathematical analysis. In this investigation, I examine the combined impact of reproduction, survival, and generation length for the growth of a population. Reproduction is associated to the ability to attract mates, to the size of the litter, and to the feeding and defending of the offspring, and is primarily of importance at the transition from one generation to the next. Survival is associated to the ability to catch pray and/or to avoid predators, to have good digestion, to have sensitive senses, to behave conformingly in the current social framework, and is primarily of importance during the lifetime of individual creatures. I examine by means of a mathematical analysis the impact of reproduction and survival on the growth of population size for different generation lengths.

Suppose that we have a population of which the fraction s survives every year. Then, if the generation time is g, the fraction of the initial number of the population at the end of a generation is s^g. We next suppose that the level of reproduction at the end of each generation will increase or decrease the size of the population by the factor r.

At the onset of a new generation the population size has thereby, due to the two factors r and s, changed to rs^g. The number of such changes after time t is the number of generations during this time, which is t/g. Then the total number of individuals N in the population after time t will be

$$N/N_0 = (rs^g)^{t/g} \tag{1}$$

Where N_0 denotes the size of the population at the veginning of the studied period. We examine the impact on population size for different values of the three variables r, s, and g. Calculations are performed by means of numerical examples for four different cases.

Case 1. $rs^g > 1$; The population increases.
Case 2. $rs^g = 1$; The population is stable.
Case 3. $1 < r < s^{-g}$; The population decreases.
Case 4. $r < 1$; The population decreases.

I calculate the growth of the population N/N_0 by means of formula (1). The values of the parameters r and s are chosen for clarifying illustrations of three of the four cases (Case 2 is trivial since the size of the population is constant). In each case the growth of the population over generations is calculated for two different generation lengths and the result is illustrated in Figure 1.

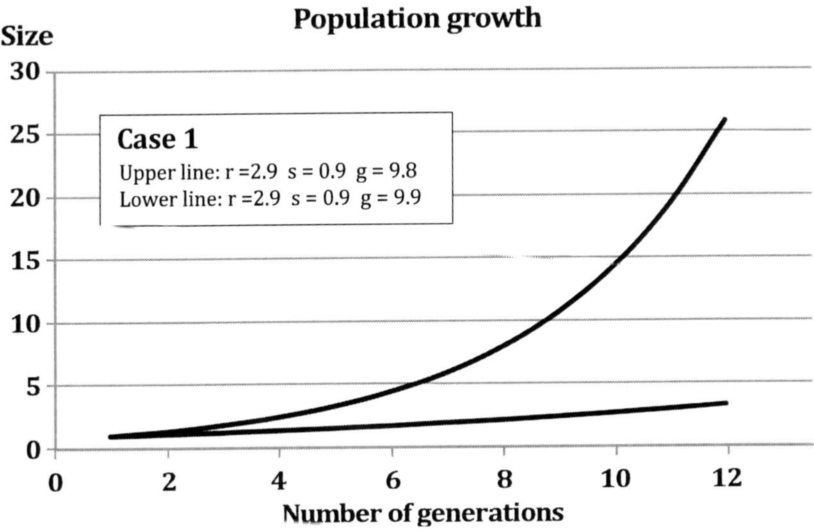

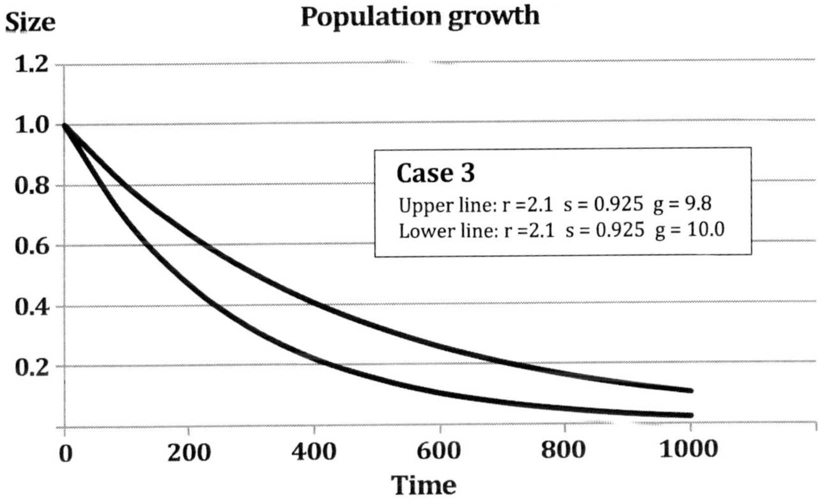

Figure 1. (Continued).

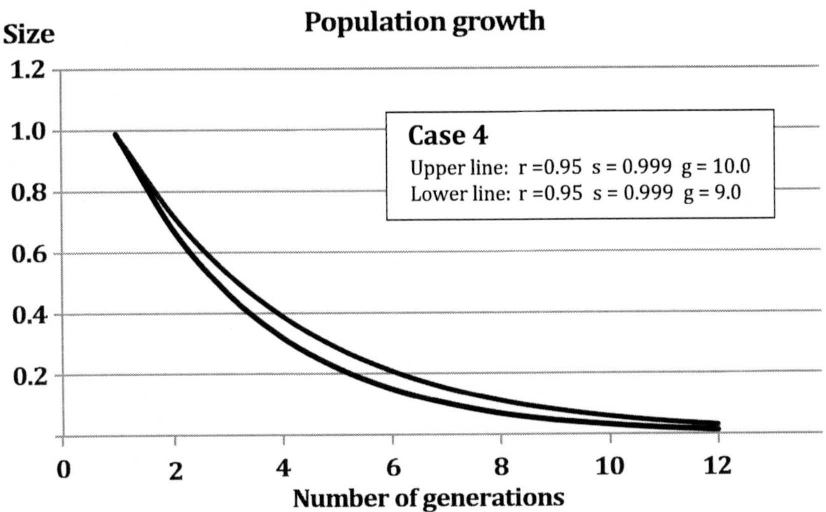

Figure 1. The graphs of Equation 1 shown for some illustrative values of the variables *r*, *s*, and *g*.

This result is understood in the following way. The annual survival factor *s*, since always less than one, reduces the population, a reduction that is compensated for at the occasions of reproduction with the factor *r* at the end of generation time *g*.

In Case 1, the graphs show that the size of the population will grow faster for a shorter generation time. In this case the environment is beneficial for population growth implying that the total reduction at the end of each generation time is exceeded by the reproduction and if generation time is shortened, the number of annual reductions falls and the number of occasion of reproduction increases, both factors of which contribute to more rapid growth of the population size over time. This means that a shortening of generation time is beneficial for the growth of the population and hence there is a selection for shortening of generation time.

In Case 2, the reproduction constant at the end of generation time balances the annual reductions and the population is stable. In this case shortening of generation time is neither beneficial nor deleterious for the population. Yet, in this case the three parameters must have very special values for satisfying these conditions, which seems quite improbable, especially as the features are stretching over time. Therefore, due to the natural variation of *r* and *s*, the conditions will very soon change to either Case 1 or Case 3.

In Case 3, the graphs of the formula show that the size of the population will decrease slower for a shorter generation time. In this case the reproduction at the end of generation time cannot fully compensate for the sum of annual reductions and the population size will decrease over time. But if generation time is shorter, the number of the annual reductions falls and the occasions of reproduction will be more numerous. Since the reproduction factor is bigger than 1, both these circumstances result in slower decrease of population size as testified by the graphs. Therefore, even under these harsh circumstances, a shortening of generation time is beneficial. This is an especially important implication of the mathematical model that, I think, hardly could have be intuitively anticipated. It clarifies the general selection pressure for shortening of generation time even in a deleterious environment. Together with Case 1 one can conclude that there is *a selection pressure for shortening of generation time, and therefore also for condensation, whether the environment is beneficial or deleterious*. Therefore the traditional interpretation of natural selection as being coupled to the external environment seems insufficient in explaining condensation and that there is an environment-independent form of natural selection responsible for the condensation of developmental traits.

If extended over a long time, this situation of course leads to extinction, a disaster however being postponed by the slower decrease of population size. In this way the population raises its chances of survival until less perilous external conditions appear or until it by means of ordinary natural selection has got time to adapt to the hostile environment. Because extinction is a far from infrequent occurrence in the history of evolution, the proposed process of shortening of generation time, I think, isn't without importance since one may think that a species despite hard external conditions in such critical periods may have avoided extinction by means of this process.

One should here remember that in the above discussion of condensation we concluded that condensation is invisible to natural selection, at least in its traditional interpretation because, since condensation implies no change of function of a trait, natural selection has no alternatives to choose amongst. This conclusion is, as we can see, concurrent with the result of the calculation, which shows that condensation actually can be independent of the environment. Still, I think one shouldn't abandon the process of natural selection when it is about condensation, just that it is of a form that is environment-independent. Because of this independence of the haphazardly varying external conditions it seems possible, maybe even plausible, that condensation could exhibit some degree of regularity.

In Case 4, the graphs of the formula show that the size of the population will decrease faster for shorter generation time.

In this case the low value of the reproduction constant adds to the decrease coming about by the small survival factor, implying that the population size will suffer a successive decrease. But in this case, as the analysis shows, the population decreases faster in case of shorter generation time. Therefore there is a disadvantage of a shortened generation time. The situation leads to extinction if not interrupted by a change to a higher value of the reproduction constant, thus bringing the population to the conditions of some of the former cases. Therefore species having in fact survived over time cannot have been applying the conditions of the fourth case except for short periods of time and therefore this case doesn't contradict the general selection advantage of shortened generation time for normal, long living species.

The main result of the mathematical analysis shows that, when a population is increasing, there is a benefit of shortened generation time. However, even if the population is decreasing as in situations when the external environment is moderately deleterious, there is a reproductive advantage of shortened generation time as well.

Hence, as the result of the theoretical analysis shows, the selection for shortened generation time is independent of the external conditions. This means, I claim, that *there is a form of natural selection that is independent of the external environment.* One may perhaps dare to say that the tendency for shortened generation time is an inevitable characteristic of living creatures. I will in a coming section demonstrate an empirical support of this conclusion.

As we see, this result is at odds with the traditional interpretation of natural selection according to which evolutionary changes are seen as adaptations to the external environment.

However, in the case of the selection pressure for shortened generation time, I cannot identify any possible feature in the environment to which condensation could be an adaptation. Therefore, natural selection in the adaptationist interpretation isn't applicable. Instead, as I have demonstrated, condensation is explained as a result of a form of natural selection that is independent of the external environment. This conclusion is concurrent with the conclusion as posed above, implying that condensation of developmental traits occurs without change of their function, and that therefore natural selection has no alternatives among which to choose.

Richard Dawkins (2004 B, p. 88) mentions the notion of evolutionary change *per se*, and I suggest one may use this idea and state that environment-independent natural selection is a process that, as opposed to adaptive change, might be characterized as evolutionary change *per se*. I think that this general notion can be meaningfully used in a comprehensive description of evolution.

An Application to Bacteria

The result of the analysis in this section is primarily of theoretical interest though I can think of an application regarding the growth of bacterial colonies. In a high-nutrient environment a population of bacteria increases and there is a selection for shorter generation time according to Case 1. In a situation of depleted nutrients or exposition to antibiotics, the population may decrease. At the application of antibiotics to pathogenic bacteria, the result of Case 3 is especially interesting, implying as it does a selection pressure for shorter generation time and a postponement of eradication under the conditions of moderate population decrease. The bacteria then have a somewhat prolonged time for adaptation to the antibiotics thus getting an opportunity of developing resistance. It is therefore important that the antibiotic treatment is given as a high dose therapy so that the decrease of the bacterial colony is fast enough to enter Case 4. The vital importance of a complete extinction of pathogenic bacterial colonies is already well understood though I think the present result gives additional emphasis to the importance of high antibiotic therapy.

As I already have pointed put, the tenet of the present work is that evolution proceeds as a result of the continuous changes in developmental programs of living organisms. After the above analysis of the developmental process, I continue by examining the coupling of the developmental process to that of evolution.

DEVELOPMENT AND EVOLUTION

The Relationship between Development and Evolution

There has since long been noticed that vestiges of old evolutionary stages can be observed in the embryogenesis of present-day individual organisms, observations that have been subject to enduring discussions related to the idea of recapitulation (for a comprehensive survey, see Richards 1992). These discussions were intense in the second part of the nineteenth century, but since they seem not to have given any satisfactory conclusion they have faded out.

In the present work I revitalize the principle of the concept but avoid the term recapitulation because of its bad reputation.

I have been criticised for dealing with this connection, which by most biologists is considered an obsolete and abandoned idea. I must therefore try to make myself very clear, especially, as it's my conviction, that the connection gives a possible key to a deeper understanding of the evolutionary process. In addition, the connection constitutes the main empirical support of present work.

The interpretations of the observed vestiges have mostly been restricted to morphological and anatomical features within the field of biology. When I also included cultural features of the human species in the analysis, I discovered a conspicuous pattern, indicating a relationship between the developmental process of a modern individual human being and its evolutionary history. This pattern is depicted in Figure 2.

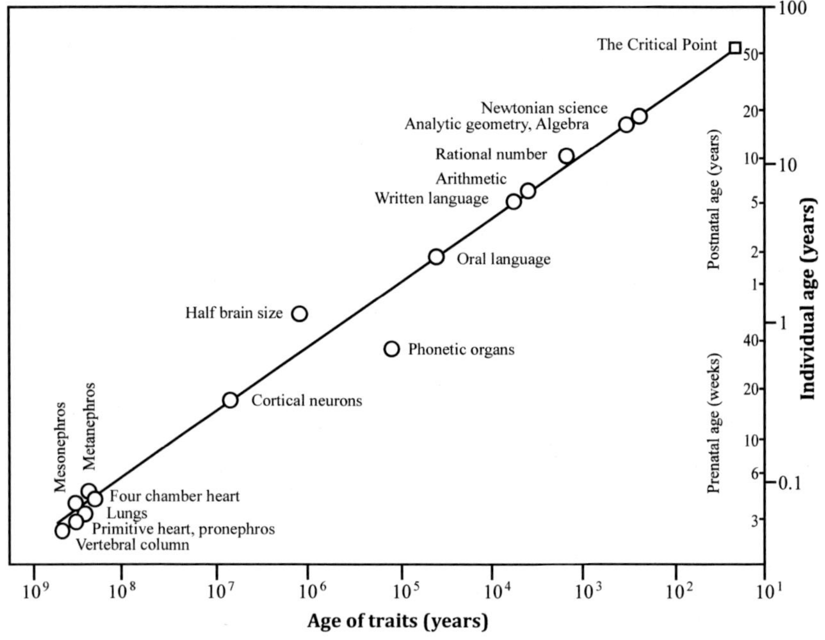

Figure 2. Individual development versus evolution of the lineage of man. The horizontal axis gives evolutionary time measured backwards from the present point of time. The vertical axis shows the individual age. Both scales are logarithmic. References to and discussion of the traits that are included in the diagram are given in my previous publication of the pattern (Ekstig 1994). This version of the diagram was published and further discussed in Ekstig (2007, 2010 A, 2015).

The pattern as revealed in the diagram of Figure 2 makes the notion of a relationship between development and evolution more concrete. As we can see, the pattern to a large extent comprises human cultural traits as well. Otherwise, the regularity would never have been disclosed and this, I think, is why biologists didn't discover it.

As we can see, the diagram reveals a concrete connection between development and evolution or, as used to be said, between ontogeny and phylogeny. Darwin himself observed and discussed this relationship as we can see in the following quotation:

> I shall attempt to show that the adult differs from its embryo, owing to variations supervening as a not early age, and being inherited at a corresponding age. The process, whilst it leaves the embryo almost unaltered, continually adds, in the course of successive generations, more and more difference to the adult. Thus the embryo comes to be left as a sort of picture, preserved by nature, of the former and less modified condition of the species. This view may be true, and yet may never be capable of proof (Darwin 1859 p. 338).

The observation of such a relationship has been subject of a vigorous though bewildering discussion during the later half of the nineteenth century. Ernst Haeckel, known as a forceful supporter of Darwin's theory, stated the phrase "ontogeny recapitulates phylogeny," also known as the biogenetic law.

One of Darwin's great contributions, besides that of natural selection, is the concept of common ancestry. In one of his notebooks from 1837, i.e., one year before his discovery of natural selection, one can find a rough drawing showing this principle (Darwin 1837). It's interesting to note that it was probably not a direct observation of the phenomenon of evolution that gave him this brilliant idea but rather the study of embryos.

As Carl Zimmer (2002 pp. 316-317) points out, Darwin acquainted himself with studies of embryos of gorillas and chimpanzees and saw conspicuous similarities. Bone for bone, they are almost identical. As human embryos develop, they pass through virtually identical stages as gorillas or chimps. Only relatively late in their development do they start to diverge, taking on different proportions. These similarities, Darwin argued, were signs that apes and humans descended from some ancient common ancestor.

In looking at Darwin's principle of common ancestry from the perspective of developmental programmes, one may conclude that species having deviated from a common lineage share the same developmental course up to the stage corresponding to their divergence, a principle called *Spencer's theorem.* McKinney and McNamara (1991 p. 390) express this observation in saying that the more closely related two groups are, the more similar their ontogenies, and the later their ontogenies diverge.

Let me also refer to a remark on this issue made by Dawkins who reminds us that organisms, including human beings, form overlapping sequences of generations in which each generation exhibits a close resemblance to its progenitors. He then suggests an expressive metaphor: "Everything about a modern animal, especially its DNA, but its limbs and its heart, its brain and its breeding cycle too, can be regarded as an archive, a chronicle of its past, even if that chronicle is a palimpsest, many times overwritten." (Dawkins 2004 A p. 23) (A palimpsest is a manuscript page from a scroll or an old book that has been overwritten and used again.) In this metaphor, the DNA has a special weight because, as we know, the DNA is copied over and over again, with only small modifications, over many generations.

What makes the evolutionary process so complicated is a double causality. Evolution proceeds as a result of modifications in the development programs of living organisms and the development course recapitulates traits that are consecutively added during the evolutionary history. This mutual causality has come into being at the construction of the straight line in Figure 2.

When considering the diagram in Figure 2, I would like to emphasize that the traits forming the pattern are the same, whether appearing in an ancient ancestor or in a modern organism. Thus for instance, the three forms of kidneys, pronephros, mesonephros, and metanephros, are identical, at least as seen to their function, in the ancient reptile and in a modern human being. Of course, we cannot observe the ancient reptile but, as is generally observed, the embryonic organs don't change much over evolutionary time, and we may therefore conclude that in to-day's reptiles, the kidneys are the same as they were in the ancient reptiles. This is supposed to be valid for all traits in the diagram. By means of this reasoning, I conclude that the similarities giving rise to the biogenetic law can scarcely be said to be similarities at all because instead, with respect to their function, they are identical. Consequently, there is one and only one point for each trait in the diagram, though given two separate dates for its appearance.

What really is enigmatic, though, is the existence of such a regular pattern over such a great part of the evolutionary process, actually over all the time of animal evolution. Of course, this regularity insists upon an explication, especially as the process of evolution is supposed to be driven by natural selection depending on the highly irregular external contingencies that certainly have been prevailing all over this long time. Step by step, I will develop an explication of this conundrum.

The straight line invites to two ways of interpretation. The first concerns condensation and the second concerns complexity.

We first discuss condensation, the assessment of which is obtained by means of a mathematical analysis of the line.

The Regularity of Condensation

The straight line of course invites to a mathematical analysis. The equation of line is

$$\ln t_o = C_1 - C_2 \ln t_p \tag{2}$$

where t_o and t_p denote the developmental (ontogenetic) and evolutionary (phylogenetic) age of each trait, respectively. The values of dimensionless constants are determined by the scales on the axes yielding $C_1 = 5.12$ and $C_2 = 0.39$.

Due to the fact that the evolutionary timescale in the diagram of Figure 2 starts at the present point of time and displays time in the negative direction, the position of each particular event will be displaced to the left as time proceeds and, because of the fact that the time scale is logarithmic, this displacement per unite time as seen in the diagram will appear smaller the higher the age of the trait. Then the line wouldn't be linear except for the present point of time, which is a highly unsatisfactory feature. Instead it is reasonable to assume that the line reflects a general characteristic of life and not just an accidental coincidence of the present point of time, because, of course, our understanding of the evolutionary process cannot rely on the presumption that the present point of time has a special status in the evolutionary process.

However, there is a possibility that the linearity of the pattern is preserved over time if one assumes that the points are displaced downwards towards earlier developmental appearance simultaneously as they are moving to the left due to the increasing age of each trait with time. In accepting this, as I claim, highly reasonable assumption, one gets not only a means to explain the straight line, but a possibility of a quantitative measure of acceleration and condensation as well. If the displacement downwards, recognized as acceleration, occurs at a pace that is determined by the value of the derivative $a = dt_o/dt_p$ of Equation (2), the linearity of the straight line will be preserved over time. Differentiation of Equation (2) yields

$$a = -C_2 t_o/t_p \qquad (3)$$

The quantity a thus denotes the acceleration of each trait appearing at the ontogenetic age t_o.

Such a regular displacement of developmental traits is the result of an appropriate shortening of all preceding stages, a shortening recognized as condensation. The quantity of condensation, denoted q, is the relative difference of two values of acceleration at adjacent points on the lines. This difference is the derivative of acceleration with respect to t_o. Hence $q = da/dt_o$, and differentiation of Equation (3) yields

$$q = -(C_2 + 1)/t_p \qquad (4)$$

This analysis shows that the straight line in the diagram is coupled to a continuous and regular shortening of developmental stages, i.e., condensation. The formula implies a systematic decrease of condensation over time, which is intuitively sensible since the more a stage is shortened, the more difficult it must be to shorten it still more. However, I find it remarkable that condensation obeys such a simple formula, depending as it does on one variable exclusively – the age of the trait.

Maybe, though, that the simple regularly of condensation, and hence of the presence of the straight line, isn't so remarkable after all, because, as we found in a previous section, condensation is a consequence of a selection process that is independent of the irregular external contingencies.

As we can see, the formula implies the possibility of a quantitative assessment of condensation and a couple of examples may clarify the significance of the formula. Since the traits of the human embryo emerged very early on the evolutionary time scale, in fact several hundred million years ago, the condensation of these traits is quite small, actually of the order of 3% over 10 million years. As a second example, we may choose the acquisition of oral language of a modern child, which, according to the model, is condensed 3% per millennium. Such small values are of no practical significance. The values of condensation of traits appearing at higher age are bigger. Thus, for a modern child, the acquisition of Newtonian science is found to be 5% per decade. In the present work, though, there is no need to use these specific values of condensation, because the suggested model of evolution is merely outlined in a most general form.

It may seem, though, that condensation has little influence on evolution at large. I can imagine a process of evolution without condensation, proceeding in much the same way that it actually has gone through.

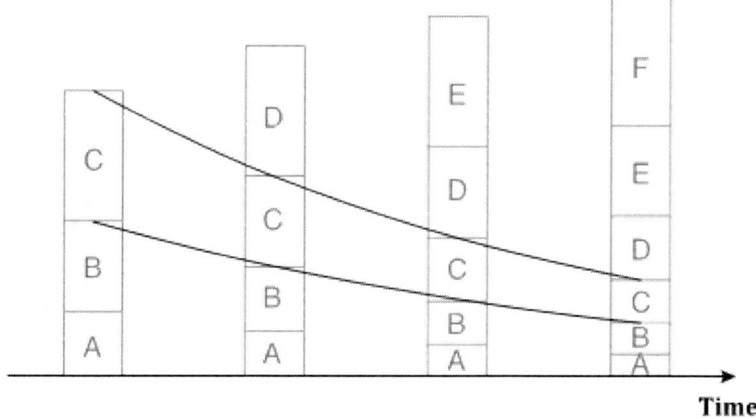

Figure 3. The vertical columns illustrate four life histories at different points of time in the evolutionary history of a species at occasions when terminal additions have occurred. The blocks in the columns represent developmental traits. We may think block C as representing a primitive back-bone, block D a primitive heart, E the lungs, and F the modern heart. It may also be suggested that the blocks represent stages in the human cultural evolution. Thus block C may represent oral language, D written language, E arithmetic and F algebra. Between these columns one may think of a lot of intermediate life histories in which there has been a continuous condensation and shortening of generation length. The curved lines illustrate acceleration and the distance between them at each point of time corresponds to condensation.

Generation times would just be longer. However, one must assume a random variation of generation times. Then, as discussed in the above section of population biology, a population with somewhat shorter generation time would reproduce more effectively and win the competition with populations with longer generation times. This is how natural selection works, in this case actually in a form that is independent of external conditions. I thus conclude that condensation is an inseparable part of the evolutionary process.

Condensation and Terminal Addition

It is important to clear out the how of the counteracting processes of condensation and terminal addition are working. I therefore try to make it easy to grasp with a highly idealized illustration, intended to facilitate the understanding of the concepts.

At the top these columns, sexual maturation sets in, marking the end of ontogeny. The figure illustrates condensation of early appearing organs during the course of evolution. Due to condensation of preceding organs, subsequent organs are displaced towards earlier appearance, a process recognized as acceleration. The figure also illustrates how new organs are added at the end of ontogeny, a process identified as terminal addition.

As we can see in this picture, the lengthening of generation time and the often-accompanying growth of body size, being normal trends in the great perspective of evolution, don't contradict the conclusion of a simultaneous presence of condensation. Such a lengthening might however be somewhat reduced by condensation. In this way, maturation time escapes from becoming awkwardly long, but this conclusion is a consequence, not a cause, of condensation. The fact that condensation and terminal addition give a superimposed result is probably responsible for the fact that the occurrence of condensation has not been given appropriate attention in prevailing discussions of evolutionary processes.

Terminal additions (including increasing body sizes) must in most cases have been more advantageous than the disadvantages due to the consequential prolongation of generation time, because otherwise the new traits could never have come about. This means in these situations that ordinary environment-dependent natural selection in the formation of new traits is stronger than the accompanying selection for shortening of generation time. Nonetheless, it seems reasonable to assume that, during the process of evolution, the counteracting processes of condensation and terminal additions have been in action simultaneously and independently of one other.

The sequence of stages at the top of the columns illustrates what is normally considered as the evolutionary history of the species. Of course, it is unavoidable to observe that this sequence of traits is observed in the sequence of traits building each individual's life history as well. This is recognized as the contentious concept of recapitulation that I have mentioned above.

An Explication of the Regular Temporal Structure of Evolution

The straight line in Figure 2 indicates a large structure of evolution that insists on explanation. In Figure 3, the four columns illustrate in an idealized way four individual developmental courses at the occasions of terminal additions. Such major transitions are certainly occurring at highly different time intervals because they are molded by natural selection and thus dependent of the haphazardly varying external conditions. Between these occurrences one may envisage a great number of individual developmental courses (not illustrated in the diagram), incessantly exposed to selection pressure for condensation. In my analysis of population biology, I found condensation to be independent of external contingencies. Therefore it seems possible, maybe even plausible, that it could exhibit some degree of regularity. If so, the lines of acceleration in Figure 3, i.e., the lines connecting the onsets of developmental traits would be even as indicated in the figure. As a consequence, the irregular time intervals at which terminal additions have occurred in the course of evolution would be preserved as corresponding time intervals at which traits are added in subsequent individual developmental courses. This structural similarity could then be illustrated as a straight line as that in Figure 2. Since this line, formed on the basis of empirical data, in fact is found to be straight, the assumption of the regularity of condensation is confirmed. One may thus conclude that in spite of highly irregular conditions, the evolutionary process has nevertheless, under the influence of the environment-independent functioning of condensation, accomplished a large-scale regularity in its outcome as manifested as a straight line in Figure 2.

A different way to arrive at this conclusion is to assume that if condensation hadn't been independent of the environment, it would probably not have been regular and if so, the temporal structure of the evolutionary process wouldn't have been preserved in the developmental process. Then the line wouldn't have been straight.

Let us, in order to make this reasoning easier to follow, consider an example related to the features of the diagram of Figure 2. In this diagram we can observe that the temporal gap between the evolutionary occurrences of the primitive heart and the four-chamber heart is much smaller than that between, say, the appearance of cortical neurons and oral language. The same difference is observed in the timing of these traits in the individual growth process as seen on the vertical axis. This demonstrates that the temporal structure of the evolutionary process is preserved in the developmental process, thus giving additional support to the above reasoning. Of course, one should not forget that the mathematical analysis of the straight line has shown that the regularity of condensation is a consequence of the linearity of the line, but the task of the reasoning has actually been to explain the feature of the line of being straight.

The main conclusion is that, *in spite of the irregular time intervals at which changes have occurred in the course of evolution, there is a large-scale regular structure of these changes, revealed when also the developmental process is taken into account.*

This regular structure is a result of a process being coupled to a form of natural selection that is independent of the external environment.

Empirical Support

An important question is of course if the discussed selection pressure for condensation in fact has been exerted in evolutionary lineages of living creatures and if so, if such a process can empirically be studied. A fundamental difficulty hampering this question is that the embryonic development of ancestral organisms is not available for studies and therefore that longitudinal study of condensation in a lineage cannot be performed. This difficulty restrains all attempts of examinations of temporal displacements as studied in the field of heterochrony. For the present analysis, studies of the age of an embryo are the most urgent. However, as Balinsky (1975 p. 302) points out, the age of an embryo is often not known, and in animals other than mammals, the rate of development is dependent on the temperature of the environment making an assessment of the age of the embryo especially uncertain. Moreover, the size of the embryo is no true indicator of its degree of development, as the dimensions of the embryo vary to a great extent. As to the morphological peculiarities of the embryo, there are certain limitations even to this approach. Finally, as he indicates, the development of different parts or organs of the embryo is not always strictly coordinated in time.

In this situation, I think, we are restricted to look at the largest patterns of the evolutionary process. Thus, from textbooks on embryology like the one by Balinsky, we may contend that in present-day vertebrates, the most vital somatic organs (the heart, the kidneys, and the like) are moulded very early in the gestation period.

One must remember in this context that new traits in their first emergence appear as terminal additions because, as is generally assumed, novel traits appearing early in the embryogenesis are most probably affecting subsequent forms in a deleterious way.

Then, after this occurrence, condensation of this and preceding traits are causing the observed early appearance of basic embryonic and juvenile traits. Since, as we have seen, developmental traits are found to have a regular temporal dependence of the intervals at which evolutionary events have occurred, it would be interesting to find out if the very process of evolution would display some degree of regularity as well. In order to find such regularity, one needs some sort of measure of evolution itself.

This leads us to the second way of interpretation of the straight line in the diagram of Figure 2 – an interpretation in terms of complexity.

COMPLEXITY

Common Views

In the extensive literature on evolution, many authors seem to be hesitating about how to determine the direction of the evolutionary process; about how to characterize, as they say, the arrow of time. A common though intuitive notion is that the direction can be characterized as a steady increase of complexity. In the present work I maintain that the concept of complexity provides a promising way to get a deeper understanding of the very process of evolution.

A recent contribution to the literature on complexity in connection to evolution is given by Lineweaver et al. (2013), having collected contributions from several researchers in the field. A common problem, expressed by many of these authors, is the lack of a definition of complexity. But, as the editors ask, even without a definition or a way to measure it, isn't it qualitatively obvious that biological complexity has increased? Do we really need to wait for a precise definition to think about complexity and its limits? (ibid. p. 5). I find this remark highly relevant for my present analysis.

Melanie Mitchel (2009 pp. 96-111) describes many forms of complexity in specific fields such as complexity of size, entropy, information, thermodynamic depth, fractals, and degree of hierarchy. She contends that complexity is discussed in several fields of research such as in genetics, in Evo-Devo, and in studies of chaos.

As the history of science shows, the comprehension of a new concept often starts with its measurement. This practice can be seen as exemplified by Newton's measurement of force and Joule's measurement of energy. Actually, there are some fundamental concepts that cannot be defined by reference to still more fundamental concepts. I think that complexity may be such a concept. If so, it is futile to wait for a definition and instead it seems better to develop methods of its measurement and explore the explanatory consequences of such a measurement. I have in a previous publication (Ekstig 2010 A) suggested such a method that shortly will be presented in the next section.

Even if one cannot formulate a stringent definition, there is need of a description in ordinary terms of what complexity means. As an attempt at such a description, I would here like to use a simple formulation by McShea and Brandon (2010), implying that complexity means the number of parts or the amount of differentiation among parts within an individual.

Most authors express the intuitive notion that complexity has been increasing during the course of evolution. Since natural selection is generally seen as furnishing the prime mechanism of evolution, it is more or less tacitly assumed that increasing complexity has come about by the action of natural selection. But there are also attempts to explain increasing complexity even without the action of natural selection (McShea and Brandon 2010, Kauffman 1993, 2013). An objection against these attempts is raised by Lineweaver (2013 p. 7), maintaining that a definition of complexity without the involvement of natural selection merely describes an increase in entropy.

However, many authors resolutely deny the possibility of a meaningful use of the concept of complexity in the context of evolution altogether, referring as they do, to the lack of a stringent definition. It seems in this context that Robert Trivers's opinion has had a pervasive influence through his statement: "There exists no objective basis on which to elevate one species above another. Chimp and human, lizard and fungus, we have all evolved over some three billion years by the process known as natural selection" (Trivers 1976). In a certain meaning, all species can of course be said to be emanating from the very first forms of living organisms, since all have parts of their DNA conserved from these old organisms. I think that that is what Richard Dawkins has in mind when stating that all lineages have had exactly equal time to evolve since the dawn of life and he ardently discards the idea that animals could be ranked on a "higher or lower" scale (Dawkins 1992 p. 263). He is somewhat more open, though, to the concept of progress, especially as a result of arms race (Dawkins 2004 A pp. 496-497).

However, Dawkins doesn't think that anybody would deny that there has been a broad overall trend towards increased information content during the course of human evolution from our remote bacterial ancestors (Dawkins 2004 B p. 101) and then, in the same context, he refers to another scientist in suggesting that information content of a biological system is another name for its complexity (ibid p. 102), a contention that, at least, he doesn't deny. It seems to me that Dawkins here comes quite near the notion of complexity as a quantity that is increasing in evolution.

I think that Trivers's and Dawkins's opinions in this matter can be understood insofar as when expressing a notion of a change of an entity, one has to rely on some kind of quantitative assessment, a measure that, as commonly observed, is lacking in the case of complexity. Yet despite this problem, there is a widespread opinion, even as we have seen amongst scientists, to hold on to the view of a generally increasing complexity in the evolutionary process, a view that is maintained with or without the involvement of natural selection.

In the present work, I strongly adhere to the view, expressed by so many scientists, that complexity has successively been increasing during the evolutionary process and that this process mainly is a result of natural selection. However, the view that increasing complexity is driven by natural selection requires a more detailed analysis that I will come back to. But, first of all, can complexity really be measured?

The Assessment of Complexity

As a point of departure, I make the assumption that there is such a concept as complexity that has been continually increasing during the course of evolution. The intuitive view of such a process relies on how evolutionary key novelties have been appearing over time. Such an indication is discussed by Carl Sagan in the form of a picture he calls The Cosmic Calendar. In this picture he presents evolutionary key novelties, including some cultural events as well, between which there are successively shorter time intervals (Sagan 1978 pp. 14-16). In his comprehensive overview of organic evolution, Richard Dawkins (2004 A) demonstrates the same pattern and Yuval Noah Harari (2012 p. viii) provides in the same vein a timetable of biological and cultural history.

As we can see from these accounts, the novelties appear irregularly but at rapidly shortened time intervals thus harmonizing roughly with the logarithmic time scale of Figure 2. One may raise the objection that the choice of these key novelties is subjective, but it seems to me that the choices, though intuitive, are reasonable and probable shared by many people. It must be emphasized, however, that these accounts of evolution cannot be used for an assessment of the rate of evolution because to that end one also needs an estimation of the degree of change or quality of each novelty. I suggest that the concept of complexity can be used for that purpose.

I have in my construction of the diagram of diagram of Figure 2 suggested the line in the diagram to be a representation of complexity (Ekstig 2010 A). The justification of this conjecture is relying on the explanatory possibilities it may infer. I developed a numerical analysis making possible an estimation of the increasing values of complexity during the evolutionary process, a procedure that can be seen as an operational definition of complexity. I suggest this particular form of complexity to be called *evolutionary complexity.*

The interpretation of the straight line in Figure 2 as illustrating evolutionary complexity relies on two assumptions. The first is that it has been incessantly increasing during the course of evolution. This assumption is supported by the fact that many scientists agree with it as I have pointed out above. Therefore, I don't find this assumption controversial.

The second assumption is that evolutionary complexity increases during the developmental process as well. The problem with this assumption is that all information about the growth of an individual creature is present in its genome at the very beginning of the developmental process and doesn't thereafter change (with some sporadic exceptions) and therefore one could say that the information for the developmental process has been present all the time from its beginning. I think, though, that it is important to distinguish two features of the DNA. First, it contains information about how to build a body and, second, it performs the building work as well. To use a metaphor – the DNA is both the architect and the builder's workman.

As we know, the architect's plan is available from the beginning and is step-by-step accomplished by the builders. In a similar way, the building of an animal body is accomplished, at the supervision of the genes, by means of successive additions of new parts and more connections between the parts, which means an increase of complexity according to the definition of complexity that I suggested above. Therefore I conclude that evolutionary complexity is increasing during the developmental growth of an individual animal. The same reasoning can be applied to the development of our intellectual faculties, though in this case, the information and instructions are stored in the social environment.

To begin with at the formation of the procedure of measuring evolutionary complexity (Ekstig 2010 A). I made the assumption that its growth may be following an exponential function. Therefore, I assumed as a starting hypothesis that evolutionary complexity is following increasing values similar to the negative tail of an exponential function. Hence, its growth is assigned values between 0 and 1.

The next step is to construct a diagram of complexity on a linear time scale. To this end, one must take into account that the position of each point on the line in the diagram of Figure 2 is determined by two time coordinates, one for evolution and one for development.

These two time coordinates for each trait is now represented by a number of pairs of points with their time coordinates on the common linear time axis stretching backwards from the present point of time, thus forming two separate sequences. Next, these pairs of points should be distributed in the diagram according to their coordinates of complexity, up to now undetermined, which is a most crucial procedure.

To that end, one must keep in mind that every pair of these points originates from one and the same point on the line in the diagram of Figure 2, thus representing the same value of complexity. The coordinates of complexity of these points are determined by a tentative formation of an evolutionary curve that gives the most sensible shape of the developmental curve. Because the time axis of the diagram is linear, it is impossible, due to the great difference of temporal extension of the processes, to display the two curves in one and the same diagram. Nonetheless, the procedure can be executed in a numerical mode.

The result of this adjustment of the two curves to each other indicates that an exponential form of the evolutionary curve is unable to give a sensible form of the development curve. Rather, I found that complexity during the course of animal evolution is best described by means of a sequence of exponential functions with increasing bases.

This means that complexity has increased in a superexponential pace, thus growing in an extremely slow pace in early evolutionary times and much faster in the cultural realm (Ekstig 2012) (A superexponential function is a variant of an exponential function in which the base is continuously increasing). The resulting diagram from this procedure is reproduced in Figure 4 in which only the last 100 years of the evolutionary curve is shown.

I suggest that we consider this assessment as a hypothesis and examine its explanatory possibilities. As I already have pointed out, the understanding of evolution has been strongly hampered by the lack of a method for its measurement and I maintain that the suggested method for the measurement of evolutionary complexity gives a possibility to discuss the progress and direction of evolution.

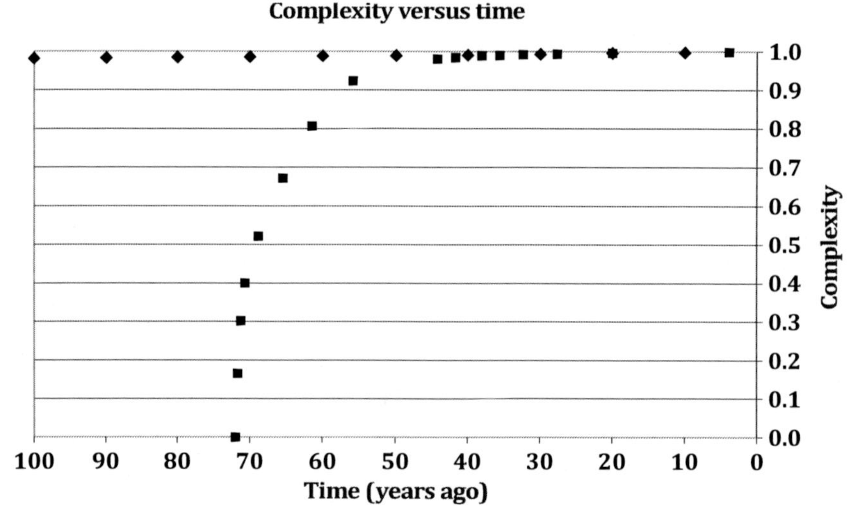

Figure 4. A combined diagram over development (squares) and evolution (diamonds) on a linear time scale. The curves are adjusted to each other to give the most sensible shape of the development curve. The evolutionary curve has its onset 600 million years ago which is a point on the linear temporal axis of this diagram, positioned 600 km to the left. At the point of time 20 years ago, the curves coincide, which corresponds to the point called The critical point in the diagram in Figure 2. The diagram is reproduced from (Ekstig 2010 A).

Table 1. Some relative values of evolutionary complexity as given by the numerical program as reproduced from Ekstig (2010 A)

Evolutionary time (years ago)	Individual age (years from conception)	Relative value of evolutionary complexity
600 million	0.06	10^{-26}
6 million	0.38	0.16
1 million	0.76 (birth)	0.30
25,000	3.2	0.52
1,200	10	0.80
400	16	0.92
100	28	0.98
20	52	0.99

The numerical analysis indicates that evolutionary complexity has been steadily increasing during the course of evolution, although the increase isn't following a simple mathematical formula. A table of its values as obtained from the suggested procedure is given in Table 1.

In the present work, though, I think there is no need of using the specific values of evolutionary complexity as received by the measurement. The measurement has its justification insofar as it makes statements about increasing complexity better grounded.

Furthermore, it makes possible the construction a diagram of complexity versus time for the evolutionary process at large, without the requirement of specifying the precise values on the axes. As it turns out, such a diagram implies an entry for several basic insights into the evolutionary process being presented and analyzed in later sections.

It is interesting to note that the evolutionary process can best be explained by means of a time scale running backwards, starting from the present point of time. Such a feature contrasts sharply to the ordinary way of studying processes in nature by following the direction of time itself, and I think it reflects the fact – by no means unnoticed – that the features of living organisms are best understood on behalf of their bygone history. This result also concurs with the fact that future routes of evolution cannot be predicted. However, the nearest future can in a limited meaning be seem as determined by natural selection. As Dawkins (2004 B p. 103) points out, natural selection is by definition a process whereby information is fed into the gene pool of the next generation. This, as I see it, explains the continuity of the evolutionary process.

Accelerating Evolution

There is a common notion that evolution is an accelerating process. According to the suggested procedure of measuring evolutionary complexity, I determined the successively increasing bases of the consecutive exponential functions representing evolutionary complexity. From these bases one can derive the doubling times of the complete process. This procedure is described and the resulting doubling times at different evolutionary epochs are given in Ekstig (2012). Here, I reproduce in Figure 5 the obtained graph of the doubling times which gives a conspicuous picture of a significant feature of evolution. Added to my own values are a couple of other measurements for the latest period of time.

If evolution were following an exponential course because of its cumulative character, the doubling time would be constant. However, the present result indicates the overall trend to be much more rapidly increasing with rapidly decreasing doubling times, implying a superexponential evolutionary course. The result implies that at about 100 million years ago the doubling time was about seven million years. Then during the following period of animal evolution it decreased to 600,000 years. At the onset of human evolution the doubling time was about 200,000 years to successively decrease to about 3,500 years at about 400 years ago, a value being characteristic for the pre-Galilean scientific evolution.

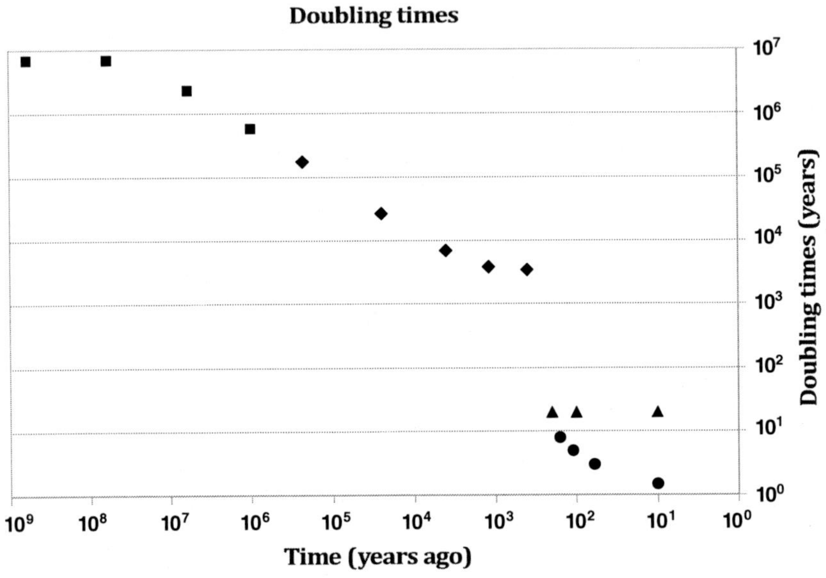

Figure 5. The doubling time of complexity versus evolutionary time. Squares represent animal evolution, diamonds hominid and human evolution including, for the last three points, pre-Galilean scientific evolution, triangles post-Galilean scientific evolution adapted from Jinha (2010), and circles the evolution of information technology adapted from Nagy (2010). The diagram is reproduced from (Ekstig 2012).

For the most recent 200 years, human scientific activity shows a dramatic progress with a doubling time of only 20 years. Finally, information technology over the last century follows a still more rapid rate with a doubling times quickly decreasing to the present value of only 18 months.

The diagram shows that evolution at large is following an accelerating pace that in the present time has reached a tremendous rate. Ray Kurzweil

(2005) has discussed this trend, suggesting that evolution soon will reach an unlimited speed, thus arriving at what he calls the singularity; a really alarming prospect. If evolution had been found to follow a hyperbolic function, one could expect it to end in such a singularity.

However, I have built the analysis of increasing complexity on a sequence of exponential functions that, although rendering the complexity a rapidly increasing pace, doesn't necessary end up in such a precarious singularity.

The concept of evolutionary complexity has hitherto in this work been discussed in descriptive terms. I now suggest an explication of its growth.

A Self-Reinforcing Feedback Process

An intrinsic feature of life is that living organisms apply the remarkable principle of starting, so to say, from scratch in the formation of each new generation. Only the genetic information is preserved, thus allowing the growing individuals to take advantage of the experiences collected by parents and previous generations. This information is then sieved by the external environment before it is fed back to the next generation or, as I have discussed above, even sieved by a selection process without the involvement of the external environment. In short, this is natural selection combined with a feedback process.

The concept of positive feedback is a source of growth of a system. It originates from the technology of electronic amplifiers, in which part of the output voltage is coupled to the input side. In this way the voltage, which I here call the substrate, is amplified by a feedback loop, a mechanism that can give a rapid and nearly unlimited degree of voltage amplification. This principle can been transferred to many kinds of processes. Of special interest in the present context is its application to the evolutionary process. Bernard Crespi (2004) points out that positive feedback can be instrumental in driving many of the most important and spectacular processes in evolutionary ecology. He emphasizes that the self-reinforcing dynamics of a positive feedback generates the conditions for changes that might otherwise be difficult or impossible for selection or other mechanisms to achieve and that it can generate large-scale changes in genetic systems.

Crespi mentions the peacock's tail, the human brain, the extinction of passenger pigeons and the infestation of our genomes by repetitive elements as examples of the many remarkable phenomena in evolution and ecology for the growth of which he suggests the mechanism of positive feedback. To his broad account of cases of positive feedback, I would like to add the implementation of complexity, which I suggest can be seen as a common concept behind Crespi's examples.

According to a common description of complexity, it means the number of parts or the amount of differentiation among parts within individuals as I have discussed above. Therefore, the addition of a new trait contributes to the complexity of the species. One may assume that this increase of complexity opens still more opportunities for the species to adapt to the environmental peculiarities and thus to add still more traits. In this way, a species is increasing its complexity according to the principle: the higher its level of complexity, the higher the number of opportunities to add still more traits and thus to increase complexity still more.

Such a procedure, even if the implication of its repetition in the developmental course is taken into account, implies a cumulative process that normally leads to an exponential increase.

Let us consider a typical example of a situation in which natural selection gives rise to a cumulative process. An animal's fur is a shelter against cold climate – the colder the climate, the thicker the fur. Thus, the climate stimulates the fur to grow cumulatively but the fur has no influence on the climate. In order that a feedback process should be applicable, there should be a coupling back from the thickened fur to the climate thus leading to a runaway process typical of a feedback in action. In this case there is no such coupling. In many cases in organic evolution natural selection works that way. Are there cases in which natural selection could give rise to a feedback process?

Let us consider such an example. *Arms race* between predators and prey implies that an improvement of the capability of one side will have influence on that of the other thus leading to runaway improvements of the capabilities of both sides. As Dawkins (2004 A p. 496) points out, arms races are deeply and inescapably progressive in a way that, for example, evolutionary accommodation to weather is not. Therefore, the mutual coupling occurring in arms race comprises, I think, the hallmark of a self-reinforcing feedback process. The improvements of the capabilities involved in this process certainly entails an increase of complexity of both sides leading to the conclusion that the self-reinforcing feedback process in arms race causes a rapid increase of complexity that may be much faster than if generated by a cumulative process only.

Sexual selection is another example of a self-reinforcing feedback process. A typical example is found in the conspicuous ornamental plumage of the peacock's tail. In this case the plumage evolution in the male is coupled to a sexual preference for such a plumage in the female. This means that an improvement of a feature on one side will have influence on that of the other. An important feature of this process, as discussed at length by Dawkins (1988 p. 203), is that the genes for male qualities, and the genes for making females prefer those qualities are evolving together leading to mutual runaway improvements, typical of a feedback process.

Another application of a self-reinforcing feedback process is found in cultural evolution in which the evolution of language is especially elucidating. This case will be discussed in a forthcoming section on cultural evolution.

Of course, such a rapidly growing process as that accomplished by the self-reinforcing feedback process cannot go on unlimitedly. The peacock's tail cannot grow infinitely. It seems that the benefit of the new trait starting the feedback process sooner or later is exhausted leading to the kind of natural selection called stabilizing selection; a situation that may go on until interrupted by the emergence of another new trait that may start a new feedback process.

It is conceded that feedback will give rise to an exponential amplification. The cases just discussed are examples of feedback amplification of complexity thus being expected to increase exponentially with time. Such an exponential increase is determined by the base of the exponential function making possible a nearly unlimited degree of amplification. However, during the extended course of organic and cultural evolution several such amplifications have occurred consecutively, each of which supposedly provided with its own characteristic base that in each case may result in a high increase of complexity over short time intervals. In this way, evolution may be characterized by a stepwise growth of complexity, a feature that will be discussed in a forthcoming section.

In addition to this process, natural selection contributes to a successive cumulative addition of new traits in the evolutionary process that may lead to a large-scale exponential increase of complexity. When also the contributions to the growth of complexity from the feedback instances are added, one may find a large-scale increase of complexity that by far exceeds a merely exponential growth.

Despite the periods of stabilizing selection, when seen in the large-scale perspective of evolution, the total pattern of increasing complexity may very well explain the observed increase of complexity as I have previously found and illustrated in Figure 5.

In this way it seems to me that feedback is an inherent principle of evolution that, together with natural selection, accomplishes the generally large-scale pattern of increasing complexity in evolution – a pattern that natural selection by itself is incapable to accomplish. It is now time to discuss complexity a bit closer.

COMPLEXITY AND THE TREE OF LIFE

Some Authors' Views of Evolution

Many renowned authors discuss the general trend of evolution. Thus John Maynard Smith, together with his co-author, Eörs Szathmáry, contend that living organisms are highly complex and that increases of complexity have depended on a small number of major transitions (Maynard Smith and Szathmáry 1995 p. 3).

Such notions indicate a stepwise increase of complexity. Edward O. Wilson, though without using the concept of complexity, expresses a comprehensive view of how life has evolved:

> Species emerge quickly and fully formed after a rapid burst of evolution, then persist almost unchanged for millions of years. And, conversely, rapid evolution is driven mostly or entirely during species formation. …The models of population genetics, the foundation of quantitative theory, predict that evolution by natural selection can be so rapid as to seem nearly instantaneous in geological time. The models also allow for stasis, or long periods with little or no evolution of a kind detectable in fossils (Wilson 1992 pp. 80, 81).

Somewhat later on he contends:

> All contemporary dynastic successions taken together present a complex and strikingly beautiful pattern across the surface of the earth. Now the comparison is to a palimpsest, an ancient parchment on which the current dominant groups are boldly spread and past rulers survive as faded traces in spaces between the lines, in shrunken niches. Mammals, the dominant large vertebrates on the land today, are accompanied by turtles and crocodilians, among the last survivors of yesteryear's ruling

reptiles. Forests of flowering plants shelter scattered ferns and cycads, remnants of the prevailing vegetation of the Age of Reptiles (ibid. p. 86).

It is interesting to note that Dawkins in his analysis of evolution uses the same metaphor of a palimpsest (Dawkins 2004 A p. 23). This metaphor is concurrent with the present view of evolution inasmuch as new species are added while the previously existing species continue their way of living unaffected.

These instances of thoughts by these three remowned scientists have important bearings on the present analysis. Thus Maynard Smith and Szathmáry's views imply evolution to have proceeded in a stepwise way.

Likewise Wilson argues that there have been stepwise bursts of evolutionary changes between which there have been long periods with little change. I interpret Wilson's view in such a way that evolution is proceeding cumulatively while most species continue to live in the mostly unchanged niches to which they are adapted. Because Maynard Smith as well as Wilson base their statements on initiated observations, I take their words as a basic empirical support for my interpretation.

I think that these rationales can be understood in a more comprehensive way by the application of the concept of complexity and the most direct way to do so is by means of a principal diagram over complexity versus time. From now on in this work, instead of using the concept of evolutionary complexity, I use the generalized concept of complexity.

Complexity and a New View of the Evolutionary Process

I suggest that we start by illustrating the evolutionary process by means of a highly idealized diagram of complexity versus time as shown in Figure 6

I have previously developed the diagram of Figure 6 as a result of an analysis of the evolution of individual developmental courses (Ekstig 2010 B). In Ekstig (2015) I have given a complementary analysis of the diagram as seen as a result of different forms of natural selection. In the present work I have extended this analysis and suggested an explanation of the stepwise increase of complexity as a result of a joint action of natural selection and a feedback process as we have discussed in the previous section.

It is common knowledge (see for instance Reece et al. 2011, p. 527) that natural selection works in three modes – disruptive selection, stabilizing selection, and directional selection. I have analyzed increasing complexity in terms of these concepts (Ekstig 2015), and in the present work I deepen this analysis.

In Figure 3 the uppermost blocks in each column illustrate the addition of new traits to the individual's developmental course. In this process, the new trait must have some advantage of its establishment in comparison to the already existing traits, and I think such an advantage involves an addition of complexity.

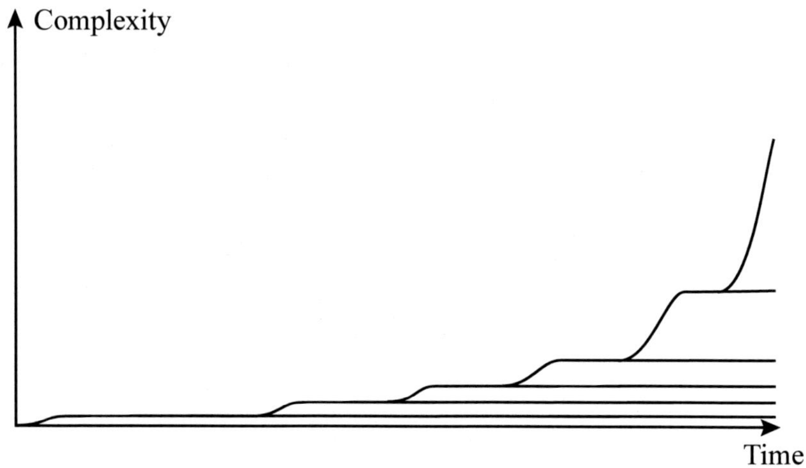

Figure 6. A view of evolution as formed by the processes suggested in the present analysis. The lines represent the complexity of animal species in a highly idealized way although the lines can be interpreted as representing the species per se as well. In this second interpretation, the diagram can be seen as a Tree of Life. The horizontal lines represent stagnant species and the stepwise curve represents the emergence of novel species at the highest degree of complexity. This stepwise curve at the same time represents the common descent of all species. The diagram was first published in (Ekstig 2010 B).

This process is in Figure 6 depicted by the steps in the uppermost line, formed by the self-reinforcing feedback process giving rise to a rapid rise of complexity in evolution, a process that is associated to the traditional concept of disruptive selection.

In the diagram of Figure 6, the stepwise raising curve thus represents species emerging in the complexity space after great though rare bursts of evolutionary transitions. After such a transition, the new species normally stabilizes at a new level of complexity, thus forming a new horizontal line above those previously formed. In this way, the diagram illustrates Wilson's comparison with a palimpsest, a metaphor for the cumulative emergence of new species above former ones. Furthermore, previously formed species are stagnant, illustrated as horizontal lines, persist to exist in what Wilson expresses as shrunken niches – an indication of the hard competition for room in the complexity space.

In order to clarify the steps in the uppermost line in the diagram, I suggest as major transitions in the biological part of evolution the emergence of multi-cellular organisms, vertebrates, terrestrial animals, mammals, and man. Other similar or less prominent transitions may be envisaged in between. After these occurrences, human cultural evolution sets in, the steps of which is discussed later on.

This procedure, when repeated over and over again, leads to a cumulative formation of new species at successive higher levels of complexity. I have suggested this rapid increase to be explained by natural selection and a self-reinforcing feedback process However, the form of the stepwise curve in Figure 6 isn't just depending on the time interval between the steps but on the height of each step as well, a feature that can be seen as a measure of the quality of the elevation of complexity at that occasion. An analysis of this feature remains to be made. In this work, I don't specify the details of the curves in Figure 6. It is in the present work sufficient to say that the stepwise curve is a principal illustration of the structure of complexity growth accomplished by the joined effects of natural selection and a self-reinforcing feedback process.

It must be emphasized that the diagram of Figure 6 is highly simplified. Thus, the vast number of species is represented by few lines only. From one such line one may imagine that a lot of related species have emerged, just with slightly diverging features on slightly different levels of complexity. In this way, the complexity space is more or less tightly filled. In the diagram these new species would form horizontal lines lying quite near those of their parent species, although such lines are not shown in the diagram. In this way one may expect that the complexity space will be tightly filled with species.

Such a situation is discussed by Conway Morris (2013 pp. 150-156), concluding that any room for exploring complexity much further is now highly restricted.

Stabilizing selection and directional selection are associated to the cases in which species are kept unchanged or with only small successive changes for long periods of time. Directional selection implies a successive change in a certain direction, for instance towards increased body size, which is a common trend in many species. Such an increased body size keeps most traits unchanged with regard to their function and implies, I think, only a slow increase of complexity. Therefore, the complexity of species applying stabilizing selection or directional selection is represented by the horizontal lines in the diagram is in Figure 6.

I summarize this section by pointing out that the rare emergences of radically new species are occurring cumulatively on the uppermost level of complexity, thus rendering the direction of the evolutionary process a significant meaning.

Competition for Room in the Complexity Space

For each of the tightly placed species forming horizontal in the diagram of Figure 6, it may be hard to change to another position in the complexity space because all nearby positions are already occupied. This contention is actually clearly expressed already by Charles Darwin in saying that "competition will generally be most severe between those forms which are most nearly related to each other" (Darwin 1859 p. 121). Daniel Dennett (1995 p. 89) has expressed a similar view, in stating that the odds are heavily against any mutation being more viable than the theme on which it is a variation. What is new in the present view is that competition is seen as occurring in the complexity space.

However, competition of this kind is not present *for species at the highest level of complexity at each point of time, because for them there are no species on higher levels causing competition.* This explains an important feature of this form of the Tree of Life implying that elevations to radically higher levels of complexity occur merely for species dwelling already at the highest level of complexity. Moreover, it gives an explanation of the direction of the evolutionary process, since these elevations to higher levels of complexity occur cumulatively. In this way, evolution becomes asymmetric – it moves mainly in one direction only, the direction of increasing complexity.

This observation is mentioned by Blum (1968 p.175) in his statement that the whole picture of evolution is one in which derivation of new patterns come from modifications of those existing – never by return to a former starting place for a new "try," unless in some very minor instances which we may neglect in our overall view.

Such an instance, though, is accomplished by parasites. These creatures are considered to be evolving towards lower levels of complexity, in the diagram diverging from a horizontal line downwards. However, they are diverging from a species that previously has raise itself to a certain higher level, and therefore parasites don't form a primary type of living organism. Hence they don't, I think, disturb the general conclusions developed in the present work. But maybe parasitism doesn't necessarily lead to a decrease of complexity. As Conway Morris (2013 p. 150) suggests, the interlocking of the genomes of hosts and parasites seems to point towards an under-appreciated degree of complexity.

The reasoning by reference to competition has a most pregnant implication. It means that the evolutionary process is going on in a continuously cumulative way, implying that old species do not suddenly jump up amongst the species dwelling on a much higher level of complexity. Such a transition would need the simultaneous change of many genetic features, which is highly improbable. This reasoning explains the simplicity of the illustration of evolution in Figure 6. The lines are not randomly pointing in different direction or crossing each other. There are no such things as hopeful monsters.

This feature of evolution is clearly demonstrated by Richard Dawkins in his great survey of biological evolution, *The Ancestors' Tale* (Dawkins 2004 A). Although Dawkins doesn't apply the concept of complexity, I find his illustrations of species evolution having much in common with the present model.

The Tree of Life

In the diagram in Figure 6, the lines represent the levels of complexity of animal species in a highly principal way, although the lines may be interpreted as representing the species per se as well, implying that the diagram can be seen as a Tree of Life. This metaphor, introduced by Darwin even before he disclosed natural selection (Darwin 1837), is now widely spread in discussions of evolution. One may find many forms of the Tree of Life in the literature. It seems that Ernst Haeckel in the nineteenth century got pervasive influence with his widely spread picture of the Tree of Life in the form of an old oak, the multitude of branching of which represents the diversity of species.

The form of the Tree of Life suggested in Figure 6 differs in important ways from the traditional form in that it includes the dimension of complexity, perpendicular to the dimension of time.

Furthermore, common descent is formed by the lineage with the highest level of complexity, a lineage represented by the stepwise curve at the uppermost part of the diagram.

Despite the highly simplified form of the Tree of Life it has, I claim, great explanatory value. It can be seen as a summary of many ideas in the present work. Especially, it explains Lamarck's challenging observation mentioned in the introduction, implying that the oldest species are the most primitive in spite of their long exposition to natural selection. The traditional solution of this conundrum is to exclude the use of the terms primitive and advanced – lower or higher – in analyses of evolution, by the same token refraining from the possibility of talking about the direction of evolution.

As I found in my previous measurement of complexity, it displays an accelerating growth. This growth is what is illustrated as the uppermost line in Figure 6. Furthermore, in the reasoning above, I concluded that the accelerating growth of complexity is caused by a combination of natural selection and repeated self-reinforcing feedback processes. However, if one refrains from the interpretation of Figure 6 as built on complexity and regards it as an illustration of the evolutionary process per se, then also the conclusions in connection to the concept of complexity may be generalized. Thus we may conclude that, although each step in the uppermost line is a result of the disruptive form of natural selection, this process is insufficient in explaining the large-scale accelerating pace of evolution. Instead, these steps are mainly caused by repeated self-reinforcing feedback processes.

The Appearance of Mankind

The cumulative addition of species with successively higher complexity implies that the latest appearing species at each point of time is the one of the highest degree of complexity. This principle has been in action all the way during organic evolution. At present, this species is the human species. As I already have commented, the vast diversity of animal species fills the complexity space very tightly. However, due to the enigmatic extinction of all hominids, there is a broad empty region in the complexity space between modern man and all other animals. This gap makes it easier to regard the human species as different because the borderline between animals and man needn't being drawn in a continuum. It is thought-provoking to fancy about the situation if the Australopithecus species or the Neanderthals were still living – by no means an improbable state of affairs. Then it had been a much greater ethical problem to determine which species could be enslaved or be used as food. In the actual state of things, this problem has vanished in that the human species is distinctly separated from the rest of all animals. Harari (2012) discusses this situation in detail.

The question of whether the human species actually can be seen as the one with the highest evolutionary complexity is a highly contentious issue and many authors seem to avoid to comment it. In his book, *The Ancestors' Tale*, Richard Dawkins starts his journey to our ancestors with a point of departure in the human species and he seems to be very careful of how to justify this choice. Thus his justification of his chosen perspective goes like this:

> Instead of treating evolution as aimed towards us, we *choose* modern *Homo sapiens* as our arbitrary, but forgivably preferred, starting point for our reverse chronology. We choose this route, out of all possible routes to the past, because we are curious about our own great grancestors (italics in the original) (Dawkins 2004 A p. 13).

I think that Dawkins's choice can be understood by use of the concept of complexity. If, for instance, he had chosen to start with the hippopotamus, his diagrams wouldn't have been so simple. I think that his choice to build his diagrams on the human lineage gives the simplest diagram, but I claim that it is only by means of the inclusion of a measure of evolution, here suggested to be complexity, that his choice is justified. Dawkins's many detailed pictures of the evolution of species are, except for this important difference, quite analogous to my picture in Figure 6.

CULTURAL EVOLUTION

Universal Darwinism

As we have seen, the preceding analyses have been based on the discovery of a regular pattern over the last 600 million years of evolution on our planet. The pattern includes the most recent time as well, thus including human cultural and scientific evolution. Indeed, these parts comprise a great deal of the pattern. Therefore, it is interesting to see in what way the discussed principles of biological evolution are applicable to culture as well.

Ever since Darwin, there have been attempts of extension of his original ideas to the field of cultural evolution.

Most of these notions have been built on analogies. A great step was taken by Dawkins (1976) with his introduction of the concept of memes as equivalence to genes. Evolution as based on memes is in this conception considered to work by the same principles as that based on genes. Thus, first there is variation so that not all forms of life are identical; undeniably true for cultural forms. Second, there is a selection that chooses those variants that have the highest ability to be spread in the prevailing environment. Third, there is some kind of heredity mechanism through which features of creatures and cultural forms are transmitted, a heredity that in the realm of culture is upheld by memes.

These conditions are thoroughly discussed by Dawkins (1976 and in many other of his works), by Dennett (1995), and by Susan Blackmore (1999), just to mention a few contributors to the extensive literature on cultural evolution in its connection to evolution.

In the above discussion of organic evolution, I emphasized the importance of self-reinforcing feedback in complexity as a central cause of the large-scale structure of evolution. Such a feedback is observed in the field of cultural evolution as well. Thus Richard Alexander (1989) proposes that social competition between and within human groups has implied a feedback of intelligence constantly producing successively enhanced human intelligence. Needless to say, intelligence is a feature of high complexity contents, implying that this feedback can be seen as working on the substrate of complexity.

We will now examine in what way the present model of evolution, as represented in Figure 6, is a good model for cultural evolution as well.

Steps of Cultural Evolution

Figure 6 illustrates the main features of the steady growth of complexity in organic evolution, characterized by major transitions in species at the highest level of complexity, and by the great diversity of stagnant species. The transitions are illustrated as steps in the graph. We now examine to what extent this model is applicable to cultural evolution.

Zimmer (2002 p. 317) suggests that human evolution is marked by five great transitions, the first of which is the separation from the apes appearing some 5 million years ago, pushing our ancestors out onto the African savannahs. The second was the invention of stone tools about 2,5 million years ago and the third the appearance of hand axes. Some half a million years ago, our ancestors learned to master fire and developed a better ability at making spears and other tools. Finally Zimmer mentions the signs of truly modern minds – paintings on cave walls, jewellery, weapons, and elaborate burials. All these transitions, I think, are good examples of increasing complexity.

Harari (2012) discusses the reasons of the human success and especially emphasizes the significance of the acquisition of language as a unique transition event of the evolutionary emergence of the human species. In the present context, I think one must consider the acquisition of language as an important step in the growth of complexity, because it certainly assumes a nervous system of high complexity. It laid the ground for the uniqueness of the human species and our conspicuous cultural faculties. I suggest the inclusion of some more recent events in our account of important evolutionary transitions, events that are included in the diagram of Figure 2.

The domestication of plants and animals, first appearing around 11,500 years ago, opened the option of abandoning the hunter-gatherer way of living and instead to live in permanent villages or cities. With this way of living followed the possibility of personal possessions and of differentiated tasks in the community. Of special importance was the invention of written language, first appearing around 5000 years ago, that I claim to be an especially important step in the growth of complexity.

I suggest that the next decisive step foreword in this summarized historic review was taken in the ancient Greek society. One may especially connect this society to the rise of the ability of arithmetic and geometry. However, after this period, religious forces kept human thought and culture imprisoned for some 1000 years during the medieval period. By the onset of the Copernican revolution, mankind took the next great leap leading to the scientific era starting in western countries and now thriving in a rapidly accelerating pace nearly all over the world.

In this way, cultural evolution follows an analogous pattern as that of species evolution, being depicted by the suggested form of a Tree of Life characterized by a stepwise cumulative increase of complexity and an accelerating rate as seen in the decreasing time intervals between the steps. As to the horizontal lines in the diagram, they are interpreted as the level of complexity in societies that by some reason or another not have participated in the growth of science and technology.

A principal difference between organic and cultural evolution is that as soon as a species is split, no reunion is possible and no exchange of experiences is practiced whereas different cultures carry out far-reaching mutual influences. Such mutual influences certainly contribute in a decisive way to the rapid evolution of human culture.

The phenomenon of culture is of course a much more intricate concept than to be represented by a few simple lines in a diagram. And it is indeed highly controversial to suggest different societies to be related to different levels of complexity. In this context I would like to refer to Jared Diamond (1997). In his attempt to explain Eurasian hegemony throughout history he argues that the gaps in power and technology between human societies do not reflect cultural or racial differences, but rather originate in environmental differences.

I would like to highlight Diamond's rationales in that it is not fundamentally a question of ethnical features that makes a distinction between different societies – these differences are by and large due to accidental environmental conditions, especially, as Diamond emphasizes, due to the advantageous occurrence of plants and animals possible to domesticate. After this crucial step, the differences between different societies have been powerfully amplified by various self-reinforcing feedback.

Analogies between Biological and Cultural Evolution

My discovery of the regular relationship between development and evolution includes human cultural traits as an important part. Otherwise, the regularity would never have been disclosed and this is, I think, why biologists never discovered it. The diagram of Figure 2 demonstrates how the cultural part of evolution has been united with organic evolution in an integrated, regular, and continuous pattern. As a matter of fact, if we accept the concept of complexity to be a meaningful measure of evolution, the model indicates that human culture provides about half of the total amount of complexity built up on this planet, as demonstrated in Table 1 – by no means an unreasonable result.

Regarding the interpretation of the Tree of Life as depicted in Figure 6 in the realm of human cultural evolution, it is important to differentiate the construction of the human body from our cultural features. The human species can in this context be regarded as an animal, being depicted by a horizontal line in Figure 6, thus indicating that our somatic features have much in common with animals and are changing only slowly. Our cultural manifestations on the other hand, as seen as expressions of complexity, are growing at an extremely rapid pace as compared with the pace of organic evolution. It should be noted that this reasoning doesn't mean that biological evolution has ceased to be in action. It is just that biological evolution proceeds so slowly as compared to cultural evolution that its effects to a large extent are hidden.

As I have pointed out, the diagram in Figure 6 initially depicts the degree of complexity of species. I also suggested that the lines in the diagram could be interpreted as representing species per se, thus forming a Tree of Life. When it comes to the cultural part of the evolutionary process this interpretation must be modified. I suggest that the lines can only be interpreted as representing the degree of complexity of specific cultural manifestations such as the faculties of verbal language, written language, and scientific advancements. However, I would like to suggest that the interpretation of the diagram shouldn't just be restricted to a descriptive demonstration of analogies but being interpreted as an indication of underlying mechanisms as well, especially the question of the application of natural selection.

Natural Selection in Cultural Evolution

In the study of cultural evolution, one may ask to what extent natural selection is applicable to this process. As to this issue, Sir Karl Popper is known to have sharply pointed out the analogy between scientific progress and genetic evolution. He suggested science may be regarded as a means used by the human species to adapt itself to the environment and asserts a fundamental similarity between the three levels of adaptation: genetic adaptation, behavioural learning, and scientific discovery (Popper 1972).

Within sociobiology, as developed by Edward O. Wilson, many human behavioural and social patterns are considered to be evolved through natural selection. Especially, Wilson concludes that gene-culture coevolution is a special extension of the more general process of evolution by natural selection (Wilson 1998 p. 128). Wilson emphasizes that culture and hence the unique qualities of the human species will make complete sense only when linked in causal explanation to the natural sciences, biology in particular (ibid. p. 267). However, I think that the role of natural selection remains to be discussed in more detail.

The Process of Heredity in Cultural Evolution

In his thought-provoking book *The Selfish Gene*, Richard Dawkins (1976) suggests cultural evolution to be analogous to biological evolution, primarily inasmuch as there is a corresponding kind of hereditary principle, the notion of memes, in the field of cultural evolution. Thus for instance, it has been asked if religion in human groups has evolved due to an assumed advantage for survival of the group. This might be an open question, but the main reason, as Dawkins expresses it, for a cultural trait to have evolved in the way that it has, is simply because it is advantageous to itself (ibid. p. 214).

When applying the concept of competition as discussed in the explanation of the diagram of Figure 6, I think one may conclude that competition is a driving force for cultural change as well. Thus Dawkins distinctly expresses such a notion: "If a meme is to dominate the attention of a human brain, it must do so at the expense of 'rival' memes (ibid. p. 211). Daniel Dennett articulates a similar view: "Minds are in limited supply, and each mind has a limited capacity for memes, and hence there is a considerable competition among memes for entry into as many minds as possible" (Dennett 1995 p. 349). It is easy to observe that, for instance, there is an incessant competition between different religions and sects, which means a competition of space in people's brains.

Complexity in Cultural Evolution

Regarding biological evolution, I have suggested the concept of complexity to have high explanatory power. We will now discuss if the concept of complexity is applicable to culture as well and if it has increased with time. Many authors intuitively anticipate the notion of a high degree of complexity in the human cultural evolution. Thus John Tyler Bonner points out that instead of all neurons and their connections being specified in the genome, only certain general parameters are specified. In this fashion, it became possible to produce nervous-system structures far more complex than anything the genes might have specified in detail. These structures are the prerequisites for human abilities in learning, teaching, and communicating (Bonner 1988 p. 192).

As to the increase of complexity, Ray Kurzweil (2005) has emphasized the rapid growth of complexity in human evolution. He regards technological evolution as an extension of organic and cultural evolution. He points out that each paradigm develops through three faces in an S-shaped curve and that evolution, both in its biological and technological expression, evolves through a series of such S-shaped curves forming a soft stair-like curve, a view in precise concordance with the diagram in Figure 6.

The Evolution of Verbal Language

As I already have pointed out, the acquisition of language forms a decisive transition event laying the ground for the emergence of the human species. I think that this event marks the transition from natural selection of the traditional form to memetic selection of the form suggested by Dawkins – both these forms of selection being superimposed during this event to be successively dominated by memetic selection. This is so because this form of selection, driven by the immaterial units of memes, results in a mode of evolution that is enormously much faster than natural selection working on the inheritance of the material units of genes. Such a great difference of the rate of change is perfectly depicted and understood by means of the logarithmic time scale allowing for a superposition of the two processes.

Susan Blackmore (1999) claims that it is the ability to imitate that makes the human species different. She assumes that people will both preferentially copy and preferentially mate with people with the best memes – especially in the case of the best language (ibid. p. 104). This is sexual selection, which, as I have emphasized above, implies a source of strong increase of complexity. Since mating is coupled to genetic evolution whereas imitation is related to a memetic evolution, the first steps of verbal evolution initiated, one could say, the transition from genetic to memetic evolution. It must be pointed out, though, that not all biologists, not to say all linguists, accept the notion of human culture as a product of Darwinian evolution. Especially regarding language, the renowned linguist Noam Chomsky is known to fervently disagree with such an idea. However, Daniel Dennett rejects Chomsky's approach (Dennett 1995 p. 390).

The question to what degree the evolution of language is driven by natural selection is a controversial question, discussed at length for instance by Dennett (1995). It seems plausible that the ability to acquire language is promoted by high intelligence and cleverness, features that certainly had high survival value in the days when all kinds of environmental hazards constantly threatened the survival of the small bands of humans. In such situations, language has certainly been beneficial for survival. But is this process really Darwinian? I think that the survival isn't coupled to a particular physical environment because the causal relation between environment and the acquisition of language isn't as apparent as that for instance between a cold climate and a thick fur. One may say that the verbal ability is equally beneficial in any physical environment. But if the social environment is taken into account, the situation may be quite different, because what is important in the social environment in this context is actually the verbal ability of the population.

Therefore, there is a kind of mutual causal coupling between the individual verbal ability and that of the population. When a child growths up it will contribute to the verbal level of its social environment that enhances the learning of language in the next generation of children. This situation means of a type of self-reinforcing feedback mechanism that I think has had a strong effect on the evolution of language. We may recall that the feedback principle is discussed above in connection with the organic part of evolution.

Thus, in the case of the evolution of language there is a kind of mutual coupling inasmuch as the child, when grown up, has impact on its environment that has consequence for the next generation of children. This process really involves the hallmark of self-reinforcing feedback: the same causal loop repeating generation after generation. This feedback mechanism, I claim, has had a strong effect on the evolution of language and has caused its rapid changes in comparison to organic changes.

There may be a more general application of the action of feedback in the evolution of the human brain. Thus Daniel Dennett (1991) concedes that "the haven all memes depend on reaching is the human mind, but the human mind is itself an artifact created when memes restructure a human brain in order to make it a better habitat for memes." Then he continues: "The memes enhance each others' opportunities: the meme for education, for instance, is a meme that reinforces the very process of meme-implantation" (ibid. p.207). Such mutual couplings, typical for feedback, have, I think, had a decisive importance for the extremely rapid growth of the human brain as seen in comparison with other somatic traits.

Condensation and Terminal Addition

The increase of verbal ability in a society is coupled to an enhancement of children's verbal skillfulness. Such an enhancement is associated to an earlier acquisition during the child's growth. In the terminology of the present work, we may conclude that there is a condensation of verbal ability. Likewise, as all of us have observed, young people continuously invent new words and a new vocabulary that, in the terminology of the present work, may be seen as terminal additions. Thus I conclude that the concepts of condensation and terminal addition, as developed in the field of organic evolution, are acting in the filed of language evolution as well.

Processes in the Evolution of Written Language

The next point after that of verbal language on the line in Figure 2 is written language. I think it is plausible that for instance in the ancient Greek society, writing was an activity exclusively performed by adult persons. I imagine the situation to be the same for the runic writing amongst the Nordic Vikings. But nowadays, most children exercise writing. This has of course to do with intentional education. I think that progress in the western society and the Eurasian hegemony to a large extent must be seen as a result of the organization of universities and, somewhat later, of school systems. During the last 200 years, there has been an awakening political awareness of the importance of education for the wellbeing of the population and that this education must be extended over merely religious indoctrination as essentially was its task in medieval times.

Moreover, education has substantially improved its pedagogical methods resulting in a rapid enhancement of learning. This also means that the children learn the specific technics at successively earlier age; in other words, there is condensation in children's learning.

It may be interesting to see in this context that even the formation of the letters has gone through a process of simplification. The gothic letters from the sixteenth century have successively been developed into more simple forms that certainly have made reading and writing more easily acquired, in this way making earlier learning possible.

In the field of evolution of written language, the changes are intentionally brought into use. This leads as to the application of intentional selection; the primary type of selection in the fields of science and technology.

Processes in Scientific Evolution

As we can see in the diagram of Figure 2, the scientific evolution forms a considerable part of the pattern. It is therefore of interest to see if one can talk about analogies between the biological, cultural and the scientific evolutionary processes as well as similarities in mechanisms.

There is a remarkably early formulation of such a notion, expressed by the 1905 Swedish Nobel laureate in chemistry, Svante Arrhenius (1859-1927) in a book from 1907, thus at the time when the discussion of Darwinian evolution was intense.

It is with ideas as with organisms. A lot of seeds are sown, but only a few will grow, and amongst the living things being evolved from them, most are weeded out through the struggle for existence, and only a few will remain living. In a similar way, ideas that are most successfully corresponding to nature are gradually selected (Arrhenius 1907. pp. 175-176, my translation from Swedish).

As we can see, Arrhenius associates the selection of ideas to the Darwinian concept of struggle for existence, in other words, to natural selection. However, there is an important difference in that, in the field of scientific evolution, the selection is intentional.

I would like to mention in passing that Arrhenius is the discoverer of the coupling between atmospheric carbon dioxide and the greenhouse effect; an issue of current interest in our own time.

The school courses in mathematics and physics, as traditionally given in many classrooms, are usually organized so that the sequence of subjects more or less parallels the historical evolution, whether this parallel is made explicit or not. This observation is in concordance with the Jean Piaget's principle of genetic epistemology.

By means of this concept, Piaget suggested that there is a parallelism between the progress made in the logical and rational organization of knowledge found in the history of science and the corresponding formative psychological processes developed in children (Piaget 1970 p. 13). Piaget reported, together with his co-author, an investigation of the relationship between the child's mental development and the cultural and scientific history, restricted to the fields of mathematics and physics. In these fields, the authors observed surprising coincidences in method as well as content and, as they emphasized, very striking common features. What is interesting about this observation is that children construct old concepts without ever being taught about them (Piaget and Garcia 1983 p. 292). Piaget's observation is especially interesting for the present model, because it emphasizes the important role of children in the process of scientific evolution. This, as we may recognize, is in line with the present model, accentuating as it does the importance of the developmental process in evolution. Especially, condensation is such a process.

The evolution of science is connected to the intentional education and the effect of school systems. These occurrences have certainly implied a change towards earlier acquisition, i.e., condensation, of the central concepts of science – indeed significant conditions for the rapid evolution of science.

Moreover, I think that the progress of scientific evolution can be accounted for by a self-reinforcing feedback process, in which the results of scientific research successively are transmitted to school and university curricula, thus enhancing the possibilities of continued progress.

I have illustrated the contribution of science to the superexponential increase of complexity in Figure 5. Of course, one must be aware of the fact that the assessment of complexity is performed by means of different methods in the different fields, and therefore conclusions of the pattern shouldn't be drawn to far. Yet, despite this limitation, the present analysis of scientific progress contributes to the suggested overarching view of evolution as an integrated process of biological, cultural and scientific evolution.

The Human Species – a Unique Species

It is, I claim, the cultural evolutionary process that contributes most to the superior complexity level of mankind. This part of the evolutionary process, according to my assessment of complexity, contributes to as much as half of the total evolutionary complexity developed on our planet. I have discussed this interpretation at length in previous publications (Ekstig 2010 A, 2015).

According to the model of evolution proposed in the present work, it is the latest appearing species that takes the position at the highest level of complexity. In our own time, this species is the human species. This gives us reason to place the human species at the highest position in the hierarchy of living creatures. This conclusion concurs with the intuitive though controversial notion of man as the summit of evolution.

In fact, this contentious notion can be traced back to the medieval idea of the Great Chain of Being, indeed even to Aristotle. It is interesting to note that this obsolete idea comprises a hierarchical but not a time dimension, whereas many scientists of today, for instance Dawkins in *The Ancestor's Tale* (Dawkins 2004 A), arrange life on a temporal but not on a hierarchical dimension. The present model, as we have seen, attempts to describe the evolutionary process by the application of both these dimensions.

The notion of mankind as a unique and superior species is a highly controversial issue not frequently expressed in scientific literature. However, a couple of examples may be mentioned. Thus, Carl Sagan eloquently emphasizes our unique intelligence:

We are a thinking species. That's what we are good at. We're not
faster than other animals, we're not better camouflaged, we don't dig
better, swim better. We think better. And because of our hands we build
better. That's our peculiar genius and the chief reason for the success of
the human species (Sagan 1989).

Likewise, Daniel Dennett in his broad survey over Darwinism boldly
states:

People ache to believe that we humans are vastly different from all
other species—and they are right! We are different. We are the only
species that has an *extra* medium of design preservation and design
communication: culture. That is an overstatement; other species have
rudiments of culture as well, and their capacity to transmit information
"behaviorally" in addition to genetically is itself an important biological
phenomenon, /.../ but these other species have not developed culture to
the takeoff point the way our species has (Dennett 1995 p. 338).

Furthermore he points out that cultural evolution operates many orders of
magnitude faster than genetic evolution, and this is part of its role in making
our species special (ibid p. 339). As we have seen, such a different rate of
evolutionary changes is a central component of the present model.

In his ambition to strengthen human dignity, American philosopher
George Kateb fervently articulates the superiority of mankind amongst all
species:

We human beings belong to a species that is what no other species is;
it is the highest species on earth—so far. /.../ All other species are more
alike than humanity is like any of them; a chimpanzee is more like an
earthworm than a human being, despite the close biological relation of
chimpanzees to human beings. The small genetic difference between
humanity and its closest relatives is actually a difference in capacity and
potentiality that is indefinitely large, which actually means that it can
never be fully measured (Kateb 2011 p. 17).

It is interesting to note that Kateb's intuitive comparison between
earthworms, chimpanzees and human beings actually coincides surprisingly
well with my own assessment of complexity, implying that the evolution of
human culture contributes with as much as half of the total value of

complexity. I just hope that my model of the evolutionary process may increase our appreciation of mankind's dignity.

Finally, in pointing out citations in favour of the notion of mankind as a unique and superior species I would like to refer to the last sentence in Dawkins's seminal book The Selfish Gene that reads:

> We, alone on earth, can rebel against the tyranny of the selfish replicators. (Dawkins 1976 p. 215).

One may wonder why so many scientists so fervently reject the depiction of animal evolution as following a direction towards successively higher levels, whereas lay people, as I have understood, see it as intuitively obvious. As I have pointed out, it seems that as to this issue, scientists refer to the current opinion that there is no measure of evolutionary progress.

Maybe also there is an apprehension that if one accepts to regard different species as living at different levels, one might as well consider different putative human races at different levels; if at all the notion of races is meaningful. I ardently emphasize that the present model of the evolutionary process doesn't endorse such an interpretation. I find support of this statement in a discussion by Dawkins (2004 B) saying that the great majority of human genetic variation is to be found within races, not between them (ibid. p. 76). This means that the horizontal lines in the diagram of Figure 6 cannot be interpreted as putative human races. This has also to do with the fact that animal species are reproductively isolated from each other whereas, as to culture, there is no such limitation. On the contrary, many human cultures have always exchanged ideas and knowledge in a way that has made their differences hard to classify. Genes are stuck to their particular species whereas memes are floating freely within and between societies.

CONCLUSION

In the present work, I give an all-embracing perspective on processes of the evolution of life and culture on earth. Thus I see the evolutionary processes as occurring in two dimensions; those of complexity and time. By means of the inclusion of complexity, evolution gets a hierarchical order.

The problem with the concept of complexity is generally seen in the absence of its definition. I have evaded this difficulty by suggesting an operational definition built on a procedure of measuring complexity by means of a combination of empirical data from the evolutionary and developmental processes. In this procedure I made use of the process of condensation of developmental traits, a process that I found to proceed regularly and independently of the haphazard external contingencies. In this way I found that the performed measurement of complexity revealed a regular and rapidly accelerating growth over a great part of the evolutionary history, explained in part as a result a self-reinforcing feedback process in complexity. In this way the wide-spread though intuitive notion of a rapidly growing pace of evolution is supported.

Natural selection is generally considered as the key process of organic evolution which, according to the common interpretation, adapts populations to their environments, thus producing enduring though irregular evolutionary changes. In the present work, I investigate complementary possibilities of an understanding of the large-scale evolutionary process.

Thus I suggest a new kind of natural selection that diverges from the traditional form in that it acts independently of environmental contingencies. This form of natural selection has its application in the process of individual development that is found to be regularly shortened. Together with the development process, the large-scale evolutionary process is found to follow a simple general pattern.

Furthermore, I have found that natural selection is insufficient in explaining the large-scale course of evolution as characterized by a rapid acceleration of its growth. This feature is instead explained as a result a self-reinforcing feedback process in complexity.

Likewise, I have found natural selection to be insufficient in explaining the human cultural and scientific manifestations of evolution. In these fields I suggest an explanation by application of two processes: a self-reinforcing feedback process and intentional selection.

My analysis of evolution in the light of complexity has led me to the construction of a new form of a Tree of Life, displaying complexity versus time. This illustration of evolution comprises stagnant species as well as the stepwise rapid bursts of novel species implying that the latest emerging species dwells at the highest level of complexity. At present this species is the human species rendering us the position as the summit of evolution on earth.

REFERENCES

Alexander, R. D. 1989. *Evolution of the human psyche*. In: P. Millar and C. Stringer (Eds.), *The human revolution: Behavioral and biological perspectives on the origins of modern humans* (pp. 455-513). Princeton: Princeton University Press.

Arrhenius, S. 1907. *Människan inför världsgåtan*. Hugo Gebers: Stockholm.

Arthur, W. 2011. *Evolution: A Developmental Approach*. Wiley-Blackwell.

Arthur, W. 2002. The emerging conceptual framework of evolutionary developmental biology. *Nature*, 415, 757-764.

Balinsky, B. I. 1975. An Introduction to Embryology. W. B. Saunders Company, Philadelphia.

Blackmore, S. 1999. The Meme Machine. Oxford: Oxford Univ. Press.

Blum, H. F. 1968. Time's arrow and evolution. Princeton University Press. Princeton, NJ.

Bonner, J. T. 1988. The Evolution of Complexity. Princeton University Press. Princeton, NJ.

Bowler, P. J. (1989) *Evolution, the History of an Idea*. Berkeley: University of California Press.

Carroll, S. B. 2005. *Endless forms most beautiful: The new science of evodevo and the making of the animal kingdom*. W. W. Norton. London.

Carroll, S. B. 2008. Evo-Devo and an Expanding Evolutionary Synthesis: A Genetic Theory sof Morphological Evolution. *Cell*. 134, 1, 25-36.

Conway Morris, S. 2013. *Life: the final frontier of complexity?* In: Lineweaver, C. et al. (2013).

Crespi, B. 2004. Vicious circles: positive feedback in major evolutionary and ecological transitions. *Trends in Ecology and Evolution*. Vol. 19 No. 12.

Darwin, C. 1837. Available at http://darwin-online.org.uk/Editorial Introductions/vanWyhe_notebooks.html.

Darwin, C. 1859. *On the Origin of Species by Means of Natural Selection*. Murray. London.

Dawkins. R. 1976. *The Selfish Gene*. Oxford: Oxford Univ. Press.

Dawkins. R. 1988. *The Blind Watchmaker*. Penguin Books, London.

Dawkins. R. 1992. *Progress*. In: Fox Keller et al. (1992).

Dawkins, R. 2004 A. *The Ancestor's Tale*. Weidenfeld and Nicolson.

Dawkins, R. 2004 B. *A Devil's chaplain*. A Mariner Book. New York.

Dennett, D. C. 1991. *Consciousness Explained*. Little, Brown and Company, Boston.

Dennett, D. C. 1995. *Darwin's Dangerous Idea.* NewYork: Simon and Schuster. p. 390.

Diamond. J. 1997. *Guns, Germs, and Steel: The Fates of Human Societies.* W. W. Norton and Company, Inc. New York.

Ekstig, B. 1994. Condensation of developmental stages and evolution. *BioScience* 44(2):158-164.

Ekstig, B. 2007. A Unifying Principle of Biological and Cultural Evolution and its Implications for the for the Future. *World Futures*, 63: 98-106.

Ekstig, B. 2010 A. Complexity and evolution: A study of the growth of complexity in organic and cultural evolution. *Foundations of Science* 15: 263-278.

Ekstig, B. 2010 B. Biological and Cultural Evolution in a Common Universal Trend of Increasing Complexity. *World Futures* 66: 435-448.

Ekstig, B. 2012. Superexponentially Accelerating Evolution. *World Futures: The Journal of Global Education* 68:1, 40-48.

Ekstig, B. 2015. Complexity, Natural Selection and the Evolution of Life and Humans. *Foundations of Science.* 20: 175-187. Open access: http://link. springer.com/article/10.1007/s10699-014-9358-y.

Fox Keller, E. and Lloyd, E. 1992. (Eds.) *Keywords in Evolutionary Biology.* Cambridge: Harvard University Press.

Gould, S. J. 1977. *Ontogeny and phylogeny.* Harvard University Press.

Gould, S. J. 2002: *The Structure of Evolutionary Theory.* Harvard College.

Hall, B. K. 2002. *Evolutionary Developmental Biology.* In: Minugh-Purvis and McNamara: Human evolution through developmental change. The John Hopkins University Press, Baltimore.

Harari, Y. N. 2012. *Sapiens. A Brief History if Mankind.* Haper Collins Publishers: New York.

Jinha, A. E. 2010. Article 50 Million: An Estimate of the Number of Scholarly *Articles in Existence.* URL: http://www.stratongina.net/files/50million ArifJinhaFinal.pdf/.

Kalinka A. T. and P. Tomancak. 2012. The evolution of early animal embryos: conservation or divergence? *Trends in Ecology and evolution*, Volume 27, Issue 7, pp. 385s-393.

Kateb, G. (2011) *Human dignity.* The Belknap Press of Harvard University Press.

Kauffman, S. 1993. *The Origins of Order.* New York. Oxford University Press.

Kauffman, S. 2013. *Evolution beyond Newton, Darwin, and entailing law: the origin of complexity in the evolving biosphere*. In: Lineweaver, Charles et al. (2013).

Kurzweil, R. 2005. *The singularity is near: When humans transcend biology*. Penguin Books Ltd.: New York.

Lineweaver, C. H., Davies, P. and Ruse, M. 2013. (Eds.) *Complexity and the Arrow of Time*. Cambridge: Cambridge Univ. Press.

Maynard Smith, J., Szathmáry, E. 1995. *The Major Transitions in Evolution*. Oxford University Press: Oxford.

McKinney, M. L. and K. J. McNamara. 1991. *Heterochrony*. New York. Plenum Press.

McNamara, K. J. 2012. Heterochrony: the evolution *Evolution: Education and Outreach*, 5(2). pp. 203-218.

McNamara, K. J. 2002. What is heterochrony? In: Minugh-Purvis and McNamara: *Human evolution through developmental change*. The John Hopkins University Press, Baltimore.

McShea, D. W. 2001. The hierarchical structure of organisms: A scale and documentation of a trend in the maximum. *Paleobiology*. 27(2).

McShea, D. and Brandon, R. 2010. *Biology's First Law*. Chicago: The University of Chicago Press.

Mitchell, M. 2009. *Complexity, a guided tour*. Oxford University Press.

Müller G. B. 2007. Evo devo: extending the evolutionary synthesis. *Nature Reviews Genetics* 8, 943-949 | doi:10.1038/nrg2219.

Nagy, B. et al. 2010. *Superexponential Long-term Trends in Information Technology*. URL: http://www.santafe.edu/~jdf/papers/Superexponential. pdf.

Piaget, J. 1970. *Understanding Causality*. W. W. Norton and Co. New York.

Piaget, J. and R. Garcia. 1983. *Psychgenèse et Histoire des Sciences*. Flammarion, Paris.

Popper, K. 1972. *Objective knowledge: An evolutionary approach*. Oxford University Press.

Raff, R. A. 1996. *The Shape of Life*. Chicago: The University of Chicago Press.

Raff, R. A. and G. A. Wray. 1989. Heterochrony: Developmental mechanisms and evolutionary results. *Journal of Evolutionary Biology*, Volume 2, Issue 6, pages 409-434.

Reece, J. B. et al. 2011. *Campbell Biology*. Boston: Pearson.

Rice S. H. 1997. The analysis of ontogenetic trajectories: when a change in size or shape is not heterochrony. *Proc. Nat. Acad. Sci.*; 94:907-12.

Richards, R. 1992. *The Meaning of Evolution*. The University of Chicago Press, Chicago.

Sagan, C. 1978. *The Dragons of Eden*. Ballantine Books, New York.

Sagan, C. 1989. "Why scientists should popularize science." *Am. J. Phys.* 57 (4).

Stearns, S. C. 1992. *The Evolution of Life Histories*. New York: Oxford University Press.

Tills, O., Rundle, S. D., Salinger, M., Haun, T., Pfenninger, M. and Spicer, J. I. 2011. A genetic basis for intraspecific differences in developmental timing? *Evol. Dev.*; 13:542-8.

Trivers, R. (1976) In: Dawkins. R. *The Selfish Gene*. Oxford: Oxford Univ. Press. p. vii.

Wilson, E. O. [1992] 1994. *The Diversity of Life*. Penguin Books.

Wilson, E. O. 1998. *Consilience*. Alfred A. Knopf, New York.

Zimmer, C. 2002. *Evolution*. The Random House Group Limited, London.

In: Natural Selection and Genetic Drift
Editor: Joshua Richardson

ISBN: 978-1-63484-331-7
© 2016 Nova Science Publishers, Inc.

Chapter 2

NATURAL SELECTION AND DIABETES MELLITUS

Svetlana Shtandel[*]

V. Danilevsky Institute for Endocrine Pathology Problems,
Kharkov, Ukraine

ABSTRACT

Advances in modern medicine enable a change in the tension of intragroup selection in human populations. Thus, implementation of insulin for type 1 diabetes mellitus (DM) treatment considerably lowered the selection tension for this symptom and converted it from the sub-lethal to the one with a lowered adaptability. Increasing variety of type 1 DM and type 2 DM is being observed recently in different populations. Moreover, recently the heterogeneity of type 1 DM and type 2 DM has also been observed. The investigation was aimed to study the influence of the selection on the evolution of DM clinical forms. Global implementation of insulin therapy into type 1 DM treatment caused the prevalent increase of this disease. Currently, there is a positive selection trend for type 2 DM, which is the original cause for the prevalence increase within the population, and the negative one for type 1 DM determines its prevalence within the population approximately on the same level. Intra-population change of gene frequencies, susceptible for type 1 and 2 DM, predetermined the development of such DM clinical

---

[*] E-mail address: shtandel@mail.ru.

forms as LADA (latent autoimmune diabetes in adults). It also resulted in the increasing number of patients with an absolute insulin deficiency of type 2 DM, which is a more complicated form of this DM type. Polymorphisms association, changing the immune response and forming the susceptibility to type 1 (C1858T of the gene *PTPN22*, A49G of the gene *CTLA4*) and type 2 (E23K of the gene *KNJ11* participating in insulin insufficiency formation) with different DM forms illustrates the result of decreasing the selection tension against type 1 DM after insulin therapy implementation into the public health services practice.

Keywords: type 1 and 2 diabetes mellitus, latent autoimmune diabetes of adults, natural selection, relative adaptability, prevalence in population

INTRODUCTION

Population dynamics factors that change the gene frequencies include: a) recurrent mutations; b) natural selection; c) migration and d) casual fluctuations [1]. It is known, that the genes and chromosomes mutations are the unique source of all genetic variability, although frequency of their occurrence is very low. As the mutations process is very slow, they by themselves change the population genetic structure with a very low speed [2]. Selection is the principle driving force of the evolution. Modern definition of selection is generally used as a differential reproduction of various genes variants, i.e., carriers of certain traits, have more chances to survive and have geniture, than the carriers of other traits [2]. The central concept of the selection theory is adaptability. Only the different reproduction speed of individuals with various genotypes is important for selection. The reproductive ability of a particular genotype in comparison with the norm is called the Darwinian adaptability of this genotype [1]. The adaptability level depends on the decrease of an individual viability and because of those or other pathologies, impossibility to survive till the reproductive age, and also the fertility decrease of any individual [3].

I. I. Shmalgauzen, who has contributed greatly into the problem of evolution studying and search of its solving, wrote: "As phenotypes are the carriers of viability and objects of the natural selection – the individual development course could not bur influence the evolution... the most important thing - an organism as such with its active struggle for its life is not visible in the genetic theory of natural selection." And, according to I.I. Shmalgauzen, phenotypes are the objects of natural selection, or, it can be said, that natural selection influences the phenotypes. Purpose of the thesis about selection influence on phenotypes assumes the necessity of relation between the selection and the organism properties [4].

Natural selection concept is a discordant problem of evolutionary human genetics. Despite the popularity of a hypothesis of "neutral evolution," most of the scientists believe that selection has played the principle role in evolution of species and has generated all the biological diversity of human populations [5]. While considering the large human populations, the following types of selection are distinguished: 1) intragroup, based on inter-individual adaptability differences (differential reproduction of genotypes) and 2) intergroup, which takes into consideration the differences in average adaptability of populations (differential natural growth of particular groups) [6]. There is a couple of difficulties, caused by applying the biological adaptability concept in respect to a human, measured by number of viable geniture, which does not consider the human society social organization specificity. The modern medicine allows to "correct" the phenotype manifestations of some hereditary pathologies, to create the adaptive environment for the genotypes, which in more severe constraints would be eliminated by selection, and in thus to increase their adaptability. This phenomenon was called "a dysgenic effect of medicine" [7, 1]. Change of the intragroup tension - for example, by successful treatment of multifactorial diseases (MFD) - leads to change of the distribution curve of the disease susceptibility and, hence, to its distribution increase. However, the quantitative estimation of such changes is complicated.

In the modern literature, the issues of selection in human populations [7-9] were actively studied. For this purpose, the methodology of selection intensity estimation (I_t) and its components, bound with differential pre-reproductive mortality (I_m) and differential fertility (I_f), is applied (developed by J. Crow [10]). However, meaning of selection in human populations is being disputed till now. Thus, some authors assume, that selection in the modern urbanized populations is practically absent [11], the others - that available data do not demonstrate the selection weakening in the modern urban populations [12]. Up to half of human gene pool is not reproduced in the succeeding generation because of the embryos death, fetal deaths, neonatal mortality, death-rate before reproductive age, celibacy and sterile marriages in developed countries [13, 14]. Factors of economic progress and social development of the society influence the selection orientation and intensity in populations, but do not stop its action. In different populations the components of differential fertility and differential mortality considerably differ from each other [8, 15-17].

It should be noted that on different stages of mankind history, selection intensity and orientation were exposed to considerable fluctuations. Ancient populations of hunters and collectors were characterized by a moderate level of pre-reproductive mortalities (I_m = 0.58) and fertility (5-6 descendants; I_f = 0.34), selection intensity (I_t = 1.11), which were not changed significantly during the time [6]. Selection intensifying was noticed in populations of the ancient farmers when the population density increased. Maximal values of Crow indexes, known from all published data, were observed at early stages of urbanization. In the end of XIX century in the Polish community of Pittsburgh (USA) the differential mortality component reached 2.98 (75% of children did not survive till the reproductive age), the fertility component reached 0.99, and the index of total selection was 6.92. In Moscow at the same time, the pre-reproductive mortality reached 60.0%, and the index of total selection was 3.15 [6, 8, 13]. Such high values of pre-reproductive mortalities were explained by a high mortality from infectious diseases, high population density, absence of medical aids and severe life conditions of the majority of citizens. Only in XX century, as a result of social progress and successes of public health services, there was a sharp reduction of pre-reproductive mortalities in the developed countries. Simultaneously this process was accompanied by birth rate decreasing [6, 16]. Growth of population genetic burden is a predicted consequence of the selection intensity depression [6, 18]. Numerous studies, devoted to neonatal mortality rising, increasing of individuals part with various congenital defects of development and immune system, mental and cardiovascular diseases were confirmed this thesis [12].

Particularly, thanks to development of medical practice, in economically developed countries, selection by elimination of nonviable individuals moves towards the early stages of ontogenesis. However, the biological selection on the later stages of ontogenesis in these countries is going on, though not by means of elimination, but in the form of retardation. After all, it is obvious, that self-preservation means both individual surviving, and surviving in the sequence of generations. Thus, for realization of biological selection it is enough that some individuals survive more successfully, producing more geniture than others [19].

One of the forms of selection action is elimination. On the average, in humans approximately 50% of zygotes are eliminated, 15% of embryos till birth and near 5.0% of children dye at birth, about 3% more do not survive till maturity [20, 21]. Biological selection by elimination takes place also at the later ontogenetic stages, especially in underdeveloped countries, in which at the centuries border there lived 4.4 billion persons, 3/5 of which lived without elementary hygienic conditions, and 1/5 had no access to modern medicine [21]. In such conditions, biological selection by elimination is unviable (for example, selection on the infections resistance).

From everything we know about selection in modern society, one can make two conclusions: 1) Biological selection in human population really reacts by means of selective elimination and inhibition. 2) However, in society there is no biological selection as an independent, determinative factor of evolution, irrespective of any social patterns. With appearance of labor, the biological selection is not getting weaker, but changing its final orientation, submitting to the social factors and social production. Finally, the social production realities define competition and selection course, elimination and inhibition in the modern society, i.e., are expressed as social selection which is carried out on "basis" of biological selection and consequently has biological consequences for society, biological "equivalents" [22]. Depending on the social production development level, the biological selection acts either via elimination at all stages of ontogenesis, or more generally, in the form of inhibition. Negative consequences of selection weakening through elimination, which are expressed in the economically developed countries in accumulation of "genetic burden," are corrected by the social production: development of medical practice [23] and "incoming of fresh blood" by means of immigration.

One of the basic achievements in biology is a new clear recognition, that the following is required for all biological traits: a) to explain precisely how the trait "works"; to explain from the evolutionary point of view, for what the trait exists [24]. Medical researches concentrated on features of the organism functioning and on immediate factors, explaining why in some people disease occurs, and in others - not. The Darwinian medicine states the other, evolutionary question. Why every organism, in more or less extent is susceptible for diseases? For what appendix and wisdom teeth exist? Why our coronal arteries are so narrow? Why the mammary gland cancer occurs so often now? From the first sight, the answer seems to be simple. Natural selection is a random process; therefore it cannot lead any trait to the highest perfection [25, 26]. However, the latest, more careful analysis revealed some other reasons, caused by evolution, of the organism remaining vulnerable to diseases: new environmental factors, our organism is not adapted for; the compromise construction, which makes us more vulnerable to diseases, but induces the general benefit, microorganisms, which evolve faster, than people do, and also the defensive traits, such as pain and cough, which are similar to illness, but actually they are the protective mechanisms, generated in the course of natural selection. There is a number of works, studying how the evolutionary approach creates the base for MFD understanding [27, 28].

It is known that the food consumption was irregular at early stages of human development. In those conditions the possibility to form deposits of fat tissue and carbohydrates in the case of their big consumption was an important adaptive mechanism, allowing to survive in the periods of poor nutrition and starvation. It is also essential, that at the first stages of human development, human's life was characterized by an exact allocation of men and women roles. Men went for hunting; women stored the hearth side, gave birth and brought up children. Thus, men were supposed to have stronger muscles that caused the primary development of the muscular tissue. For geniture feeding, women had to be adapted to keep energy reserves, i.e., to have developed fatty tissue that served as a protection of newborns against starvation at the time of poor nutrition. During later periods, food obtaining was carried on no longer by hunting, but by development of agriculture and cattle breeding, that also demanded great physical expenses. There was no abundance of nutrition for the majority of people. The certain balance between caloric intake and energy consumption was maintained. In a very short time period, from the point of view of evolution, the major part of mankind started to get plentiful nutrition without starvation periods and muscular energy expenses, related to severe life conditions. One can assume that a modern human, and firstly a man, who usually has less developed fatty tissue in comparison to a woman, obtains ischemic heart disease and myocardial infarction as a payment for satiety in combination with his muscular energy small demand [29].

Wide distribution of a particular genotype in a population takes place in the case when this genotype gives to its owner selective advantages against the individuals not possessing it, for example, the ability to survive in extreme conditions, in particular in conditions of hunger. In conditions of constant food excess, insulin apparatus is continually under increased loading due to changes in balance between caloric intake and energy consumption. As per J.V. Neel theory, as it was mentioned above, it is presupposed [28], that there are special genes (thrifty genes), allowing a person to use effectively the limited food resources, that leads to DM development in conditions of nutrition abundance. This theory has found its confirmation in works of W. Knowler et al. [30] and P. Zimmet et al. [31], who showed the rapid increase of 2 type DM frequency among the indigenous population of Northern America and Pacific Islands in conditions of urbanization and nutrition abundance.

Globalization processes, existing in modern society, lead to increase of outbreeding, panmixia, genes mixing, and segregation burden increase.

Significant part of a genome is invariable, monomorphic and conservative. This genome fraction is vitally important and any mutation in it is eliminated by natural selection. Along with this, the variable part of a genome is functionally less important. Therefore, variability, so-called genetic polymorphism, is the minor variability, connected with the adaptation processes and accommodation to certain environmental conditions. Levels of individual heterozygosis in representatives of the same or different populations can considerably differ [32]. In total, the whole genome includes about 4 million of such substitutions (polymorphisms). About 2.5 million of polymorphisms account for sense, coding part of a genome. Genetic polymorphisms spectra depend on geographical conditions, diet, racial (ethnic) belonging, etc. and they are provoked by the natural selection. In certain conditions, they can contribute to development of specific diseases or, on the contrary, inhibit them [33].

Thus, the problem of selection influence on MFD prevalence variation and evolution of its clinical forms is not studied enough and leaves a great number of questions opened.

The modern level of medicine allows to "correct" the phenotypical manifestations of many hereditary pathologies and create an adaptive environment for genotypes, which, because of any adequate therapy absence, would be eliminated by the natural selection, and, thus, to increase their adaptability. In due time, this phenomenon has been called "the dysgenic effect of medicine" [1, 7]. Selection of the polygenic diseases belongs to the intragroup selection. Change of the intragroup selection tension (for example, by successful MFD treatment) causes a change of the distribution curve of susceptibility to this disease and, consequently, increases its prevalence. However, the quantitative estimation of such changes is complicated. From the point of view of population genetics, the hereditary diseases are divided into diseases, associated with impaired reproductive ability, and diseases, where the reproductive ability is not impaired, either because of the defects insignificance, or because they are manifested only after the reproductive period completion. Frequency and prevalence of the first group of diseases are determined by frequency of the corresponding mutations occurrence. The other situation takes place in case of diseases, not affecting the reproduction [1]. This group includes the most widespread MFD, such as DM, schizophrenia, cardiovascular diseases, obesity etc. However, till now, the selection of diseases, not affecting the reproduction, in the conditions of "the dysgenic effect of medicine" is practically not being studied, and the major part of works is devoted to either the intergroup selection, or the inborn hereditary pathologies selection [34-36].

DM both 1, and 2 types are polygenic MFD [37-42]. According to J.V. Neel hypothesis [28], type 2 DM occurs as a result of special genes existence (thrifty genes), allowing men to use effectively limited food resources, which lead to the disease development in the abundance of food conditions. Thus, if this was the case, type 1 DM was a sub-lethal trait before the insulin therapy introduction into the health care practice. This type of DM is an autoimmune disease, in the course of which the acute lymphocytic insult leads to pancreatic β-cells destruction with subsequent development of the absolute insulin insufficiency [43]. Beginning at pre-pubertal or pubertal age, not supported by the insulin therapy, resulted with the lethal outcome approximately in 0.5-1 year after manifestation. Thus, possibility to leave ay geniture was practically reduced to zero. Global implementation of the insulin therapy at the end of 40th – beginning of 50th years into the health care practice has prolonged patients' life, allowed to keep geniture, which resulted in weakening of the selection tension and led to the disease prevalence increase within the population.

Nowadays, a rapid growth of type 2 DM prevalence [44] becomes perceptible worldwide. Besides the environmental factors, promoting growth of this disease prevalence, most likely, the increase of population frequency of gene complexes of susceptibility to this type of diabetes is also necessary for such increasing of type 2 DM frequency in populations. In addition, there should be noted the ubiquitous spreading of type 2 DM severe clinical variant with development of the absolute insulin insufficiency (AID), caused by depletion of residual secretory function of pancreatic β-cells and their subsequent apoptosis [45]. According to the literature, approximately 40% of patients with type 2 DM require the insulin therapy (IT) [46]. There was shown an increased family accumulation of type 2 DM in patients with type 2 DM with AID, and the genetic analysis results demonstrated that the long non-insulin-dependent (LSNID) form can formally be considered as "a less burdened" one, and type 2 DM with AID - as "a more burdened" one [47].

Is it also impossible to ignore the considerable prevalence among the clinical forms of DM variants (2-12% of all the cases) of such its form as the latent autoimmune diabetes of the adults – LADA, whose manifestation is similar to type 2 DM. The subsequent slow development of autoimmune aggression against the pancreatic β-cells, insulin-dependence and necessity of the insulin therapy became the foundation for distinguishing of this disease form as the type 1 DM subtype in classification of 1999 [48]. On the basis of the investigation of patients with the determined diagnosis of type 2 DM and availability of the auto antibodies against GAD, several authors concluded, that this form of DM has an autoimmune nature, but differs from type 1 DM by the rate of the absolute insulin dependence development and by prevalence of alleles, predisposed to type 1 DM [49, 50]. Genetic analysis of this DM form, which was carried out by us, has shown its genetic independence and considerable number of common genes in determination of clinical variants of the disease course (the % of common genes with type 1 and 2 DM was 65.3 and 66.1%, respectively) [51]. Such heterogeneity of DDM clinical variants of DM course can be the consequence of the global implementation of insulin into the health care practice, which sharply reduces the selection tension against type 1 DM that led to increasing of prevalence of gene complexes of susceptibility to the 1th type of disease within the population.

The aim of the investigation was to study the selection influence on heterogeneity of DM clinical forms and the prevalence dynamics within the population.

MATERIALS AND METHODS

To accomplish the assigned tasks, the obstetric anamnesis and presence of polymorphisms of C1858T of *PTPN22*, 49A/G of *CTLA4* and E23K *KCNJ11* genes were studied in DM patients, treated in V. Danilevsky Institute for endocrine pathology problems in 1997 – 2014.

The obstetric anamnesis data of 2106 women after 45 without any symptoms of the investigated diseases were used as a control group. The obstetric anamnesis data are obtained during examinations on different enterprises in Kharkov.

To determine the relative adaptability parameters and to calculate the selection coefficients, the data of obstetric anamnesis were used. The latter included the number of childbirths, pregnancies, spontaneous abortions, extra-uterine pregnancies, survived and deceased geniture till age of 25. It was studied either by questioning or by the hospital histories of the women with completed reproductive period (aged after 45). The number of the investigated women is represented in Table 1.

The determination of C1858T polymorphism of *PTPN22* gene was carried out on 85 patients with type 1 DM, 140 patients with type 2 DM, 115 persons with LADA and 11 healthy Kharkov citizens. The data on C1858T of *PTPN22* gene polymorphism of 242 healthy Kharkov citizens and 296 patients with type 1 DM, received as a result of the inspection in 2004, was kindly submitted for the further analysis by the research worker of Institute of parasitology and biomedicine (Granada, Spain), Maria Ivanovna Fedets. The author expresses a deep gratitude to M.I. Fedets for the data provided. The analysis results were published in the papers of M.I. Fedets as a co-author. Determination of 49A/G polymorphism of *CTLA4* gene was carried out on 64 patients with type 1 DM, 127 patients with type 2 DM, 109 persons with LADA and 75 healthy Kharkov citizens. Determination of E23K polymorphism of *KCNJ11* gene was conducted on 47 patients with type 1 DM, 101 patients with type 2 DM, 83 persons with LADA and 44 healthy Kharkov citizens. Characteristic of the investigated persons is presented in the Table 2.

**Table 1. Characteristics of the women with
the investigated obstetric anamnesis**

	Control	Type 1 DM	Type 2 DM
N	2106	281	538
Age (years)	57.47 ± 0.16	52.52 ± 0.53	60.62 ± 0.46
Age of the DM beginning (years)	-	20.79 ± 1.84	43.51 ± 0.50
BMI, kg/m^2	24.08 ± 0.12	23.58 ± 0.36	30.64 ± 0.60
WHR	0.76 ± 0.02	0.71 ± 0.06	0.90 ± 0.01

**Table 2. Characteristics of persons, whose C1858T of *PTPN22*, 49A/G of
CTLA4 and E23K of *KCNJ11* genes polymorphisms were analyzed**

	Control	Type 1 DM	Type 2 DM	LADA
Age (years)	35.40 ± 0.90	37.10 ± 0.70	53.94 ± 0.47	52.19 ± 1.45
Age of the DM beginning (years)	-	23.15 ± 0.94	44.53 ± 0.51	45.80 ± 1.34
Age of the insulin therapy beginning (years)	-	23.15 ± 0.94	53.14 ± 0.81	48.06 ± 1.37
Duration of effective per oral anti-hyperglycemic therapy (years)	-	-	10.48 ± 0.43	2.57 ± 0.32

The "accumulated" population frequency of type 1 and 2 DM in Kharkov region was calculated on the basis of data about patient's age at DM beginning and a proband age, contained in 1303 histories (857 – type 2 DM, 446 – type 1 DM), available in the district hospitals, as well information about the overall population of the district [52].

Sampling was formed according to the initial visit in 2007. The indexes of «accumulated» population frequency of type 1 and 2 DM of 1973 were estimated based on the data about age of DM beginning from 682 clinical records (402 type 2 DM and 280 – type 1 DM), received from archive of Kharkov antigoiter dispensary during 1973. The data from 213 hospital histories, which were also received from archive of Kharkov antigoiter dispensary, were used for calculation of this index for type 1 DM in 1995. The sampling was formed according to the initial visit of Kharkov antigoitrogenic clinic in 1995. The structure of the clinical variants of type 2 DM course both in 1984, and in 2007 was determined from the hospital histories data, received from the archive of Kharkov antigoiter dispensary and V. Danilevsky Institute of endocrine pathology problems. The dynamics of DM prevalence was estimated according to the official statistics data [52-58].

Relative adaptability (w) and selection coefficient (s) were calculated for selection direction determination [2]. Thus, relative adaptability (w) was estimated as a result of fertility and survival rate, and s = 1-w. The selection direction was determined as difference between selection coefficients within the population and among the patients ($\Delta s = s_{population} - s_{patients}$).

The fertility component was determined by calculation of an average geniture number per one woman, for ill and healthy women, with the subsequent division of an average geniture number per each group by a greater average geniture number in comparison groups.

Survivability component was determined by calculation of the geniture part that survived till 25 in the comparison groups with subsequent division of the survived geniture part for each group on a greater part of the survived geniture in comparison groups.

Values of «accumulated» population frequency were calculated according to [59] on the following formula:

$$q_g = q_0 + q_n \prod_{i=0}^{i=n-1} (1-q_i),$$

where q_0, q_1, q_2, ... q_n – values of morbidity accordingly in the initial (q_0), 1, 2, ...n age-dependent interval.

$$q_n = a_{n+s} p_{n+s}$$

where a_{n+s} is a part of persons, who have got ill at the age of n among the patients, whose age was n+10 years at the moment of research; $_sp_{n+s}$ is the disease prevalence among the persons of age of n+10.

DNA extraction was conducted on whole blood, using DNA extraction set "DNK-sorb-V" kit (Moscow, the Russian Federation).

C1858T polymorphism of *PTPN22* gene is determined by the polymerase chain reaction using *Rsa*I restrictase [60]. A 218-bp fragment containing the single nucleotide polymorphism C1858T of *PTPN22* gene was amplified using direct ACTGATAATGTTGCTTCAACGG and reverse TCACCAGCT TCCTCAACCAC primers. The reaction mixture contained 20 ng of genomic DNA, 1.2 µl of 10x PCR buffer, 1.2 µl of dNTP (1.25 mmol/l), 0.3 µl of each primer (20 pmol/µl), 0.6 µl of DMSO, and 0.1 µl of *Taq* polymerase ("Sibenzyme" company) in a 12-µl reaction mixture. Amplification conditions: initial denaturation for 2 min at 94°C, followed by 35 cycles of 94°C for 30 s, 30 s at 60°C, and 30 s at 72°C and an additional extension of 2 min at 72°C. Amplification product (12 µl) was incubated using 10 units of *Rsa*I restrictase ("Sibenzyme" company) at temperature 37°C for 12 h. Restricted products were electrophoresed on 2.5% agarose gel. The mutant 1858T allele loses the restriction sequence and consists of one 218 bp fragment. 1858C allele restriction results in two fragments of 176 bp and 46 bp.

A49G polymorphism of *CTLA4* gene was amplified with the help of direct 5'- GCTCTACTTCCTGAAGACCT and reverse AGTCTCACTCACCTT TGCAG primers. The reaction mixture contained 25 ng of genomic DNA, 2.5 µl of 10x PCR buffer, 0.6 µl of dNTP (2.5 mmol/l), 0.3 µl of each primer (20 pmol/l) and 0.2 µl of Taq polymerase ("Sibenzyme" company) in 25 µl of the reaction mixture. Amplification conditions: initial denaturation for 4 min at 94°C, the following steps 58°C - 45 s, 45 s at 72°C, and 45 s at 94°C (30 cycles) and the final extension for 4 min at 72°C. Amplification product (10 µl) was incubated with the restriction enzyme *Bbv*I at 65°C for 1 h. Normal allele 49 A loses the restriction sequence and consists of a 162 bp fragment, the mutant 49G allele restriction produces 88 bp and 74 bp fragments [61].

E23K polymorphism of *KCNJ11* gene was amplified with the help of direct GAATACGTCCTGACACGCCT and reverse GCCAGCTGCACAGGA AGGACAT primers. The reaction mixture contained 25 ng of genomic DNA, 2.5 µl of 10x PCR buffer, 0.6 µl of dNTP (2.5 mmol/l), 0.3 µl of each primer (20 pmol/l) and 0.2 µl of Taq polymerase ("Sibenzyme" company) in 25 µl of the reaction mixture. Amplification conditions: initial denaturation for 3 min at 95°C, the following steps 95°C – 1 min, 1 min at 62°C, and 1 min at 72°C (35 cycles) and the final extension for 5 min at 72°C. Amplification product (10 μl) was incubated with the restrictase *Ban II* at 37°C for 2 h and 65°C for 10 min. The mutant 23K allele loses the restriction sequence and consists of a 218 bp fragment. 23E allele restriction results in 178 bp and 40 bp fragments [62]. Evaluation of relative risk of the disease development at polymorphism carriage (Odds ratio) was conducted according to [63].

Statistical analysis. Data are expressed as either mean ± SD or percentages. The normality of the distribution of variables was tested with the Kolmogorov-Smirnov test. Comparison of variables between the control and diseases groups was carried out with the Student's test or the chi square test. Comparison of percentage variables was carried out with the Fisher test. A p-value equal to or less than 0.05 was considered to be statistically significant.

RESULTS

The obstetric anamnesis data, which formed a basis for calculation of relative adaptability indexes, are presented in Table 3. The compared groups of healthy and type 1 DM women differed by number of pregnancies. With the similar women behaviour related to the family planning, this fact might be an evidence of selection effect on the impregnation stage. Unfortunately, this question in the investigated groups remained opened, that does not allow to make an unambiguous conclusion about difference in influence of natural selection among healthy women and patients with type 1 DM during impregnation. Number of pregnancies in healthy women significantly exceeds those of type 1 DM, though the average number of childbirth is approximately equal within the population and the investigated groups. This phenomenon can be explained by the fact that women with this diagnosis, knowing about the problems, associated with pregnancy with type 1 DM, are doing their best to maintain the pregnancy.

The values of spontaneous abortions and extra-uterine pregnancies in the compared groups had no significant differences. The obtained results can prove that on the embryogenesis stage there was no selection influence on 1 and 2 type DM. Women distribution by number of childbirths is presented in the Table 4.

The obtained data show the significant distribution difference in childbirths number of healthy women and women with DM. Thus, the significant difference in distribution of childbirths number between women with type 1 and 2 DM is also noted ($\chi^2 - 15.319$; df = 9; p = 0.004). It is shown that number of women among patients with type 1 DM, who had two and more childbirths, was significantly higher, than of the healthy ones (χ^2 = 34.770; df = 1; p = 0.000) – 45.68 and 42.40%, respectively. Analogous result was observed also at type 2 DM, (χ^2 = 28.678; df = 1; p = 0.000) – number of type 2 DM patients and healthy women who had two and more childbirths, was 55.39 and 42.40%, accordingly. According to Table 3, patients with type 2 DM left more children, than the healthy ones and woman with type 1 DM. It should be noted, that women, who have deceased till 45, were excluded from the sampling of patients with type 1 DM because it could overstate some indexes of average number of childbirths in this group.

Table 3. Fertility indexes and geniture survival rate

Parameter	Population	Type 1 DM	Type 2 DM
Pregnancy, ($\bar{x} \pm S_{\bar{x}}$)	4.06 ± 0.07	$2.80 \pm 0.17^*$	4.01 ± 0.11
Childbirth, ($\bar{x} \pm S_{\bar{x}}$)	1.41 ± 0.02	1.47 ± 0.06	1.59 ± 0.04
Spontaneous abortions, ($\bar{x} \pm S_{\bar{x}}$)	0.08 ± 0.01	0.12 ± 0.02	0.10 ± 0.02
Extra-uterine pregnancies, ($\bar{x} \pm S_{\bar{x}}$)	0.03 ± 0.01	0.03 ± 0.01	0.03 ± 0.01
Childless women, (%)	12.35 ± 0.72	13.88 ± 2.07	10.24 ± 1.31
Children, survived till 25, (p_s)	0.972	0.877*	0.966
Children, deceased till 25 (p_s)	0.028	0.123*	0.034

Remark* - significance of differences in comparison to the population (p < 0.001).

Table 4. Women distribution by number of childbirths

Childbirth number	Control, n = 2106	Type 1 DM, n = 281	Type 2 DM, n = 538
0	260	39	55
1	953	113	185
2	735	96	243
3	125	23	44
4	17	9	8
5	6	1	0
6	4	0	1
7	4	0	1
8	0	0	1
11	1	0	0
χ^2		18.236	36.746
P		0.001	0.000

Analysis of the geniture survival rate in healthy women and mothers with type 2 DM did not demonstrate any high mortality among patients' geniture, being a little lower than in children of healthy women (Table 3). At the same time, analysis of the geniture survival rate in women with type 1 DM and healthy women registered an increased mortality among the geniture of women with type 1 DM in comparison to the children of the healthy women. The obtained data are consonant with the data on increased perinatal mortality in newborns, whose mothers had DM (without diabetes types separation), presented in literature [64, 65].

Indexes of relative adaptability and selection coefficients, calculated based on the obstetric anamnesis, are shown in Table 5. The presented results demonstrate the relative adaptability reduction in the group of patients with type 1 DM.

Table 5. Relative adaptability and selection coefficients

Group	Relative adaptability (w)			Selection coefficient (s)
	Fertility component	Survival rate component	Total adaptability	
Population	0.959	1.000	0.959	0.041
Type 1 DM	1.000	0.901	0.901*	0.099*
Population	0.887	1.000	0.887	0.113
Type 2 DM	1.000	0.988	0.988	0.012*

Remark* - significance of differences in comparison to population (p < 0.001).

Selection coefficient value in type 1 DM group is more than twice higher, than among the healthy women of Kharkov (F = 13.22; p < 0.001), that shows existence of selection against (Δs = -0.058) this DM type nowadays (Figure 1).

Due to global application of the insulin therapy into health care practice since the beginning of the 50th years of the 20th century, the selection tension against type 1 DM was reduced. For rather short period of time, it has decreased practically from 100.0% (patients left no geniture) to 9.9% that, certainly, had to affect the population frequency of the disease.

Changes of type 1 DM prevalence among the population of Kharkov area from 1973 to 2007 are presented on Figure. 2.

As expected, because of global implementation of insulin into the therapy of 1 type disease in the 50th years of the last century, approximately one generation later (from 1973 to 1995) the prevalence of type 1 DM has grown – from 0.16 to 0.32% in Kharkov region (F = 1633.47; p < 0.001). From 1995 till 2007 growth of this disease type prevalence was practically stopped and was 0.32-0.36%. And in recent years from 2009 to 2013, the decrease in population prevalence of similar disease from 0.36 to 0.26% is even noted. It can be explained by the fact that growth of genes frequencies of susceptibility to 1 type DM, caused by the insulin therapy introduction, reached its maximum and now the reduced adaptability of this trait leads to decreasing of its population frequency due to low geniture survival rate.

It should be noted, that in official statistical collections, which data were used in our work, LADA is considered as a variant of type 1 DM and number of patients with this form is summed up with number of patients with a classical type 1 DM. It can also influence the indexes of type 1 DM prevalence.

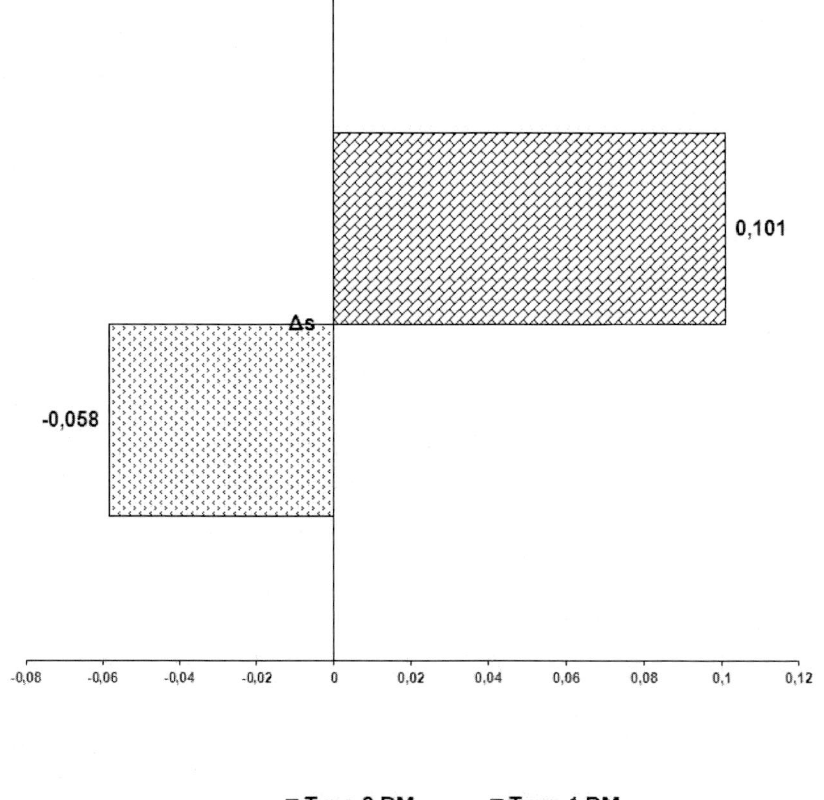

Figure 1. Selection direction of type 1 and 2 DM.

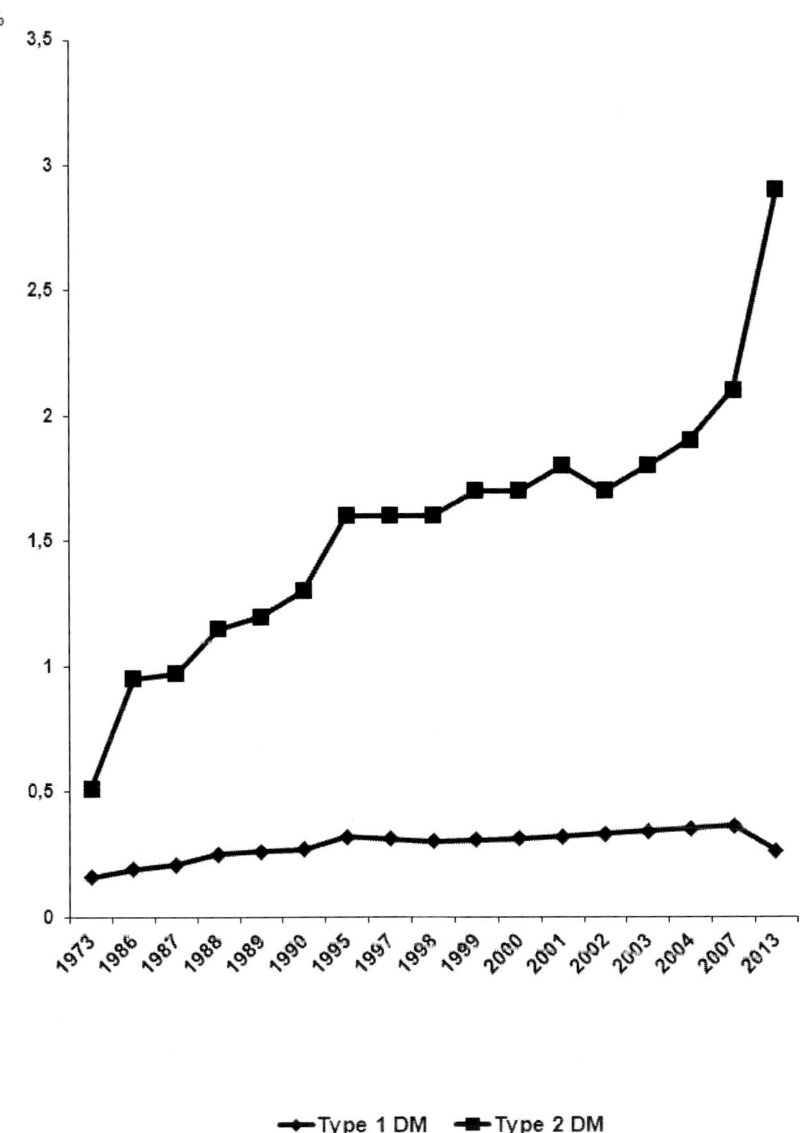

Figure 2. Type 1 and 2 DM prevalence in population of the Kharkov region in 1973 2013.

At calculation of prevalence, as the relation of patients number to the population number without accounting the demographic population structure, the prevalence increase can be caused by the population aging, because in the older age groups, as it was before, during the pre-insulin era, the previous cases are accumulated. For a more exact characteristic of dynamics of disease prevalence within the population, the age disease estimates are used, based on which «the accumulated population frequency» is calculated. To estimate the dynamics of DM occurrence probability during life, in population of Kharkov region, the indexes of «accumulated population frequency» were calculated for 1973, 1995 and 2007. The age indexes of «accumulated population frequency» of type 1 and 2 DM among the population of Kharkov region are presented on the Figure 3.

For type 1 DM, some growth of «accumulated population frequency» from 1973 to 1995 by 1.86 time is shown (F = 0.116; p > 0.05), and even its small decline from 1995 to 2007 from 0.345 to 0.343% (F = 0.0001; p > 0.05) is shown.

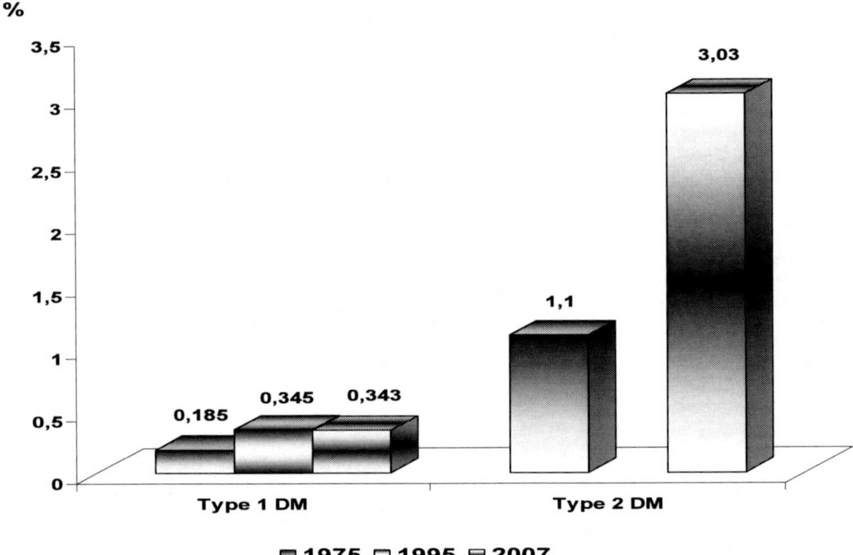

Figure 3. Indexes of «accumulated population frequency» of type 1 and 2 DM among Kharkov region citizens, %.

Existent selection against type 1 DM gives the grounds to assume that it is not necessary to expect a rapid growth of type 1 DM prevalence in populations.

Data, provided in Table 5 indicate the higher relative adaptability of patients with type 2 DM in comparison with a healthy phenotype. The value of selection coefficient in type 2 DM was much lower, than in the healthy women of Kharkov region (F = 93.06; p < 0.001). Thus, the conducted research shows that a positive orientation of type 2 DM selection (Figure 1) takes place (Δs = 0.101) – patients leave more geniture, than the healthy people, at their practically identical survival rate (Tables 3, 4). This, in its turn, increases the frequency of susceptibility to this type of disease genes in the population.

The obtained data are confirmed by the evidence on a significant increase (almost four times) of type 2 DM prevalence in population of Kharkov region (F = 31518.6; p < 0.001) for thirty four years in 1973-2007 (Figure 2.).

From 1973 to 2007 the «accumulated population frequency» of type 2 DM increased in 2.75 times (F − 5.21; p < 0.05). Therefore, the selection in favor of type 2 DM allows to forecast the further growth of this type of disease prevalence in populations and, from genetics point of view, explains the reasons, promoting a rapid growth of this type DM prevalence in the world.

The increase of frequency of both 1 and 2 type DM susceptibility genes in the population, caused by a positive selection direction (for type 1 DM to the middle of 90th years of the 20th century), can also be reason, causing the clinical heterogeneity of this disease.

Such aspect of selection action is confirmed both by the features of genetic heterogeneity of type 2 DM, and dynamics of clinical forms structure in all cases of type 2 DM. Recently, ubiquitous growth of specific severity of the disease clinical course with development of AID among all patients with type 2 DM [46] is noted. So, compared to 1984, the number of patients with AID among all patients with type 2 DM were up to 22.65% in Kharkov region, in 2007 this index has grown to 33.48% (χ^2 = 10.729; p = 0.001). The conducted researches [47] have shown that out of two studied forms of type 2 DM, one is more complicated. According to the genetic analyses results, the LSNID form can possibly be formally considered as "a less complicated," and type 2 DM with AID - as "a more complicated."

Hence, the population frequency increase of genes of susceptibility to type 2 DM, caused by a positive selection direction, raises probability of their high concentration in a proband, and thus promoting the increase of AID prevalence among all cases of the disease.

The frequency increase of genes of susceptibility to 1 type disease became the possible effect of selection tension decrease against type 1 DM, caused by introduction of insulin therapy into the health care practice. Combination of population frequency growth of genes of susceptibility to type 1 and 2 DM and increase in probability of their occurrence in a proband has defined appearance of such DM form, as LADA, among all variants of the disease course. This is evidenced by the genetic analysis results, which have showed that inheritance of LADA is described by the polygenic threshold model parameters, in its inheritance the essential role belongs to the genetic factors, there are nonlinear genetic interactions, and influence of genes number is possible with the expressed effect in determination of this DM form. Existence of a large number of common genes of susceptibility to type 1 and 2 DM and LADA (65.3 and 66.1%, respectively) defines the features of clinical course of this DM form - manifestation is similar to type 2 DM: the subsequent fast development of autoimmune aggression to the pancreatic β-cells, insulin-dependence and insulin therapy necessity [51].

Taking into account, that modern level of researches embraces not only the genetic analysis data, but also molecular genetics methods, we studied the distribution features of the single nucleotide polymorphisms genes, determining the susceptibility to both type 1 and 2 DM. Taking into consideration, that the HLA system antigens, which generally determine susceptibility to type 1 DM, are strongly linked and badly combined, it was reasonable to study the immune response genes polymorphisms, which play the secondary role in type 1 DM development, localized on different chromosomes and which are well combined in the population in the process of crossing. As an example of such polymorphisms, distributions of mutations of C1858T of *PTPN22* gene and 49A/G of *CTLA4* gene in the patients with type 1 and 2 DM, LADA and healthy Kharkov citizens were studied.

The gene *PTPN22* (protein tyrosine phosphatase non-receptor type 22) is located on the 1p13.3-p13.1 chromosome. This gene encodes the lymphoid-specific LYP phosphatase, which suppress the T-lymphocytes activation. Two variants of *PTPN22* gene - C1858 and T1858 - differ by an important part of amino-acid sequence, which is responsible for LYP association with Csk-kinase (negative regulation). When in the 620th position (T1858 allele) arginine is replaced by tryptophan, the connection with Csk-kinase is not provided. It leads to T-lymphocytes signal function disorder and promotes the development of autoimmune diseases susceptibility, in particular, to type 1 DM [66]. Today it is known, that C1858T polymorphism of *PTPN22* gene is associated with type 1 DM in many world populations: in Ukrainian [67], Italian [68], Swedish and Finnish [69].

The *CTLA4* (cytotoxic T-lymphocyte associated antigen-4) gene also defines the susceptibility to type 1 DM. It is localized on the chromosome 2q33, between two T-lymphocytes genes-activators: the receptor activator gene (CD28) and gene inducing the costimulant (ICOS), contains 4 exons and three introns. The receptor isoform (full-length isoform-flCTLA4), synthesized in the activated T-lymphocytes is coded by 4 exons: the leader protein is coded by the exon 1, the ligand-binding protein – by the exon 2, the membrane-spanning domain – by the exon 3 and the cytoplasmic domain – by the exon 4. More than 30 points of the mono-nucleotide substitutions in different areas of *CTLA4* gene are known. One of important mono-nucleotide substitutions is a single nucleotide polymorphism 49A/G in the first exon (replacement of adenine by guanine in the 49th position causes replacement of threonine by alanine in the 17th codon of amino-acid sequence of the leader peptide), leading to decrease of CTLA4 protein functional activity. It is established, that the protein product of *CTLA4* gene takes part in T-lymphocytes activity regulation and plays the important role in the autoimmune processes development [70]. Researches of different populations have revealed both availability [71-76] and absence of [77-79] the associations of type 1 DM with 49A/G polymorphism of *CTLA4* gene.

Among the single polymorphisms, determining the development of susceptibility to type 2 DM, it was logical to investigate the mutation, which is responsible for development of insulin insufficiency in a person. Polymorphism of E23K of *KCNJ11* gene belongs to such mutations. *KCNJ11* gene is localized on the 11th chromosome in area of p15.1. This gene encodes synthesis of Kir6.2 protein (Potassium inward rectifier 6.2), which is a part of the potassium channel in cells, capable to activation, and creates pores for transportation of potassium ions with the cells. The channel closing is necessary for insulin secretion by β-cells. Majority of mutations of this gene are missens-mutations, which have the dominant-negative effect, leading to decreasing of IK1 stream, reduction of repolarization and increase of action potential duration. This polymorphism takes part in the insulin insufficiency development in patients with type 2 DM and is associated with its development in various populations [80].

The comparative analysis of genotypes distribution with C1858T of *PTPN22* gene polymorphism among all compared groups (Table 6) revealed the statistically significant difference in distribution of healthy individuals and patients with LADA, type 1 and 2 DM by genotypes with C1858T of *PTPN22* gene - $\chi2 = 47.063$, p = 0.000.

Investigation of distribution of C1858T genotypes of *PTPN22* gene has shown the significant homozygotes association of this polymorphism with LADA, type 1 and 2 DM. It should be noted, that homozygous carriers of this polymorphism occur significantly more often among patient with LADA than among the ones with 1 and 2 type DM. Significant distinctions in genotypes frequencies between the patients with type 1 and 2 DM was not revealed (Table 7).

Table 6. Frequencies of genotypes alleles of C1858T polymorphism of *PTPN22* gene in DM patients and healthy Kharkov citizens

Index	Control, N = 253		Patients		χ^2	Significance of diffe-rences (p)	OR (95% CI)
	N	%	n	%			
Type 1 DM, N = 381							
Genotype: C/C	185	73.10	228	59.84	11.231	0.000	0.55 (0.55-1.09)
C/T	66	26.10	123	32.28	2.502	0.114	1.35 (0. 08-1.62)
T/T	2	0.80	30	7.87	13.674	0.000	10.73 (0.73-11.83)
Type 2 DM, N = 140							
Genotype: C/C	185	73.10	79	57.30	9.230	0.002	0.48 (0.47-1.12)
C/T	66	26.10	52	35.88	3.527	0.060	1.67 (0.80-1.95)
T/T	2	0.80	9	6.90	9.385	0.002	8.62 (0.54-11.97)
LADA, N = 115							
Genotype: C/C	185	73.10	54	47.30	21.479	0.000	0.33 (0.39-0.97)
C/T	66	26.10	43	37.30	4.092	0.043	1.69 (0.78-2.01)
T/T	2	0.80	18	15.5	30.346	0.000	23.29 (8.30-46.57)

Table 7. Distinctions in frequencies of genotypes and alleles of C1858T polymorphism of *PTPN22* gene in DM patients

Compared groups	Index	χ^2	Significance of differences (p)
Type 1 DM – type 2 DM	Genotype: C/C	0.362	0.547
	C/T	0.877	0.349
	T/T	0.150	0.698
Type 1 DM - LADA	Genotype: C/C	5.466	0.019
	C/T	0.818	0.366
	T/T	5.257	0.022
LADA – Type 2 DM	Genotype: C/C	1.906	0.167
	C/T	0.009	0.922
	T/T	4.741	0.029

Thus, the investigation of association of single nucleotide of C1858T polymorphism of *PTPN22* gene has shown its evident association with LADA, type 1 and 2 DM.

The obtained data overlap with existing works on this polymorphism investigation in populations of Ukraine, the USA, Poland, Finland, Sweden, Norway and others [67-69]. For example, in Sweden populations [69] the association of this polymorphism with all DM clinical forms was shown: frequencies of carriage of mutant homozygote T/T in patients with type 1, 2 DM, LADA and in control were 0.3, 3.8, 2.8 and 4.3% accordingly.

Table 8. The frequencies of genotypes with 49A/G *CTLA4* gene polymorphism in DM patients and healthy Kharkov citizens

Index	Control, N = 75		Patients		χ^2	Significance of differences (p)	OR (95% CI)
	N	%	n	%			
Type 1 DM, N = 64							
Genotype: A/A	29	38.67	12	18.75	5.644	0.014	0.37 (0.31-1.41)
A/G	34	45.33	29	45.31	0.028	0.866	1.00 (0.51-1.95)
G/G	12	16.00	23	35.94	6.266	0.012	2.95 (0.72-3.56)
Type 2 DM, N = 127							
Genotype: A/A	29	38.67	26	20.47	6.986	0.008	0.41 (0.36-1.28)
A/G	34	45.33	63	49.61	0.195	0.659	1.19 (0.61-1.91)
G/G	12	16.00	38	29.92	4.187	0.041	2.24 (0.69-2.93)
LADA, N = 109							
Genotype: A/A	29	38.67	22	20.18	6.681	0.010	0.40 (0.35-1.30)
A/G	34	45.33	41	37.61	0.800	0.371	0.73 (0.48-1.58)
G/G	12	16.00	46	42.20	12.943	0.000	3.83 (0.87-3.89)

The most expressed association of this polymorphism was observed with LADA. The investigation results allow to assume that development of such DM form as LADA is caused more by changes in the genes, controlling normal immunological homeostasis, and that determines the features of course of this disease form even in all genes of HLA system absence, which are necessary for type 1 DM development.

Additionally, we studied the distribution of polymorphism of 49A/G of *CTLA4* gene, which causes the immune disorders, in patients with different clinical variants of DM.

The results obtained are presented in Table 8.

These results are concordant with the data about 49A/G of *CTLA4* gene polymorphism association with type 1 DM in different populations of the world [240]. To determine the frequency of this polymorphism in different clinical variants of DM course in Kharkov population, we carried out the comparative analysis of frequencies of 49A/G *CTLA4* gene polymorphism among the patients with type 1, 2 DM and LADA.

The genotypes distribution analysis among all compared groups has shown the reliable difference in patients with LADA, type 1 and 2 DM in comparison with the healthy Kharkov citizens ($\chi2 = 17.853$; $df = 6$, $p = 0.001$).

There was shown a significant association of G/G mutant homozygotes with type 1, 2 DM and LADA ($OR_{type\ 1\ DM} = 5.61$; $OR_{type\ 2\ DM} = 4.27$; $OR_{LADA} = 7.30$). It should also be noted, that homozygous carriers of 49A/G of *CTLA4* gene polymorphism appeared more often among the patients with LADA than among the patients with type 2 DM (Table 9).

Table 9. Distinctions in frequencies of genotypes and polymorphism alleles of 49A/G *CTLA4* gene in DM patients

Compared groups	Index	χ^2	Significance of differences (p)
Type 1 DM – type 2 DM	Genotype: A/A	0.008	0.929
	A/G	0.166	0.684
	G/G	0.458	0.498
Type 1 DM - LADA	Genotype: A/A	0.001	0.975
	A/G	0.698	0.403
	G/G	0.425	0.515
LADA – Type 2 DM	Genotype: A/A	0.011	0.915
	A/G	2.953	0.086
	G/G	3.834	0.049

No significant distinctions in the genotypes frequencies in patients with type 1 and 2 DM are revealed.

The comparative analysis results of E23K polymorphism of *KCNJ11* gene distribution, participating in development of insulin deficiency and determining the type 2 DM susceptibility, among the patients with LADA, type 1 and 2 DM are given in Table 10.

Table 10. The frequencies of genotypes and alleles of E23K polymorphism of *KCNJ11* gene in DM patients and healthy Kharkov citizens

Index	Control, N = 44		Patients		χ^2	Significance of differrences (p)	OR (95% CI)
	n	%	n	%			
Type 1 DM, N = 47							
Genotype: E/E	32	72.73	8	17.02	26.410	0.000	0.08 (0.02-0.12)
E/K	9	20.45	16	34.04	1.479	0.224	2.01 (0.52-3.50)
K/K	3	6.82	23	48.94	17.743	0.000	13.10 (0.83-15.26)
Type 2 DM, N = 101							
Genotype: E/E	32	72.73	19	18.81	36.744	0.000	0.87 (0.15-1.28)
E/K	9	20.45	33	32.67	1.670	0.196	1.89 (0.57-3.06)
K/K	3	6.82	49	48.51	21.309	0.000	12.88 (2.88-15.44)
LADA, N = 83							
Genotype: E/E	32	72.73	20	24.10	26.150	0.000	0.12 (0.03-0.91)
E/K	9	20.45	20	24.10	0.059	0.808	1.23 (0.45-2.66)
K/K	3	6.82	43	51.82	23.285	0.000	14.69 (3.92-19.20)

Genotypes distribution comparison of E23K of *KCNJ11* gene among compared groups revealed significant association of polymorphic homozygote carriers with type 1, 2 DM and LADA ($OR_{1\ type\ DM}$ = 13.10; $OR_{type2\ DM}$ = 12.88; OR_{LADA} = 14.69).

It should also be noted, that the homozygous carriers of E23K of *KCNJ11* gene mutation appear more often among the patients with LADA, when comparing to the patients with type 1 and 2 DM (Table 10). There are no significant differences in genotypes frequency among the patients with type 1, 2 DM and LADA (Table 11).

Figure 4 shows distribution of polymorphisms 49A/G of *CTLA4* and C1858T of *PTPN22* genes in patients with different clinical variants of DM course.

Table 11. Distinctions in frequencies of genotypes and alleles of E23K polymorphism of *KCNJ11* gene in DM patients

Compared groups	Index	χ^2	Significance of differences (p)
Type 1 DM – type 2 DM	Genotype: E/E	0.001	0.973
	E/K	0.001	0.982
	K/K	0.017	0.897
Type 1 DM - LADA	Genotype: E/E	0.519	0.471
	E/K	1.027	0.311
	K/K	0.017	0.895
LADA – Type 2 DM	Genotype: E/E	0.478	0.489
	E/K	1.243	0.265
	K/K	0.088	0.767

The significant distinctions of distribution of genotypes with 49A/G of *CTLA4* and C1858T of *PTPN22* genes (χ^2 = 23.485; df = 16; p = 0.000) among the compared groups of patients were shown. The pair-wise comparison of DM clinical variants shown the significant distinctions between the patients with LADA and type 2 DM (χ^2 = 10.622; df = 8; p = 0.031). There was found no significant distinctions in this polymorphisms between the patients with LADA and type 1 DM; and type 1 and 2 DM (χ^2 = 4.098; df = 8; p = 0.393; χ^2 = 3.970; df = 8; p = 0.410, accordingly). The frequency of carriage of homozygote GG/TT genotypes in the patients with LADA was significantly higher than such in the patients with type 2 DM (8.30 and 1.60%, accordingly, χ^2 = 4.425, p = 0.035).

Distribution of patients by carriage of polymorphisms of 49A/G of *CTLA4* gene and E23K of *KCNJ11* gene is presented on Figure 5.

The significant distinctions in distribution of genotypes with 49A/G of *CTLA4* gene and E23K of *KCNJ11* gene (χ^2 = 19.978; df = 16; p = 0.000) among the compared groups of patients were shown. The pair-wise comparison of DM clinical variants shown the significant distinctions between the patients with LADA and type 1 DM (χ^2 = 12.065; df = 8; p = 0.017); between the patients with LADA and type 2 DM (χ^2 = 10.976; df = 8; p = 0.027). The significant distinctions in this polymorphisms between the patients with type 1 and 2 DM were not found (χ^2 = 7.859; df = 8; p = 0.097). The frequency of carriage of homozygote GG/KK genotypes in patients with LADA was significantly higher than the one in the patients with type 2 DM (21.54 and 8.86%, accordingly, χ^2 = 3.840, p = 0.050).

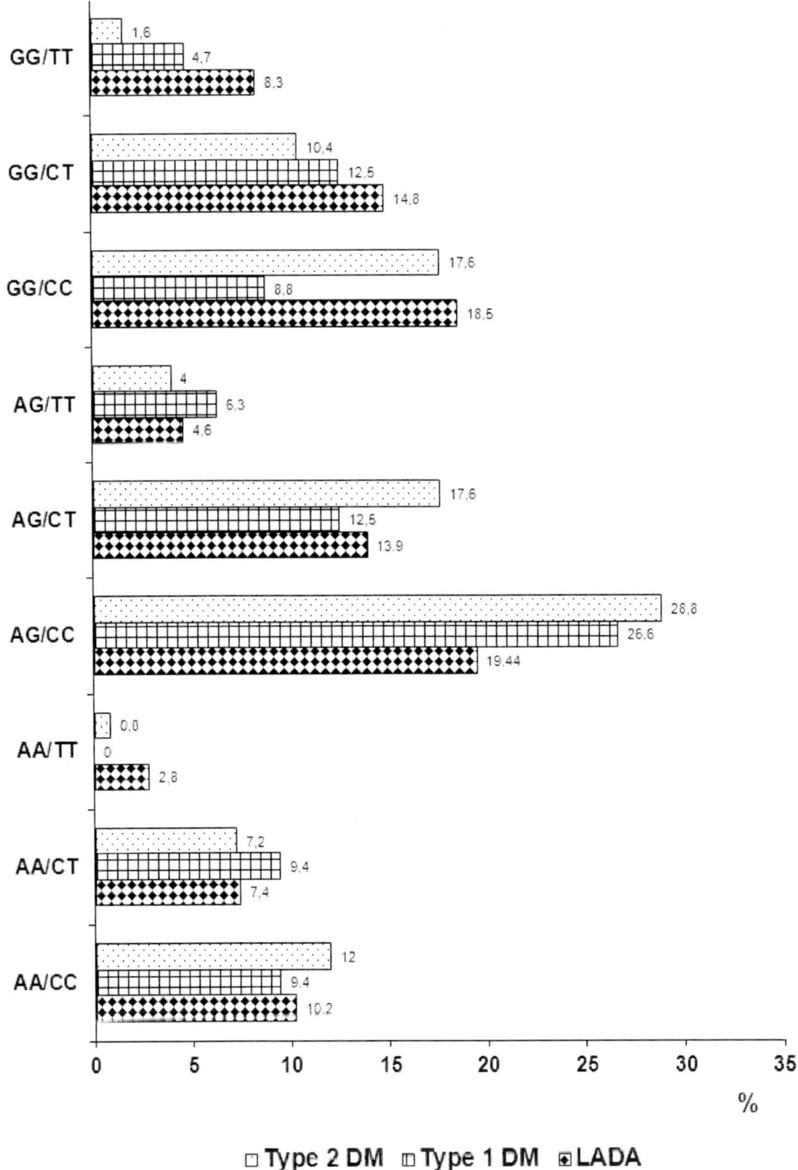

Figure 4. Distribution of polymorphisms 49A/G of *CTLA4* and C1858T of *PTPN22* genes in patients with different clinical variants of DM course.

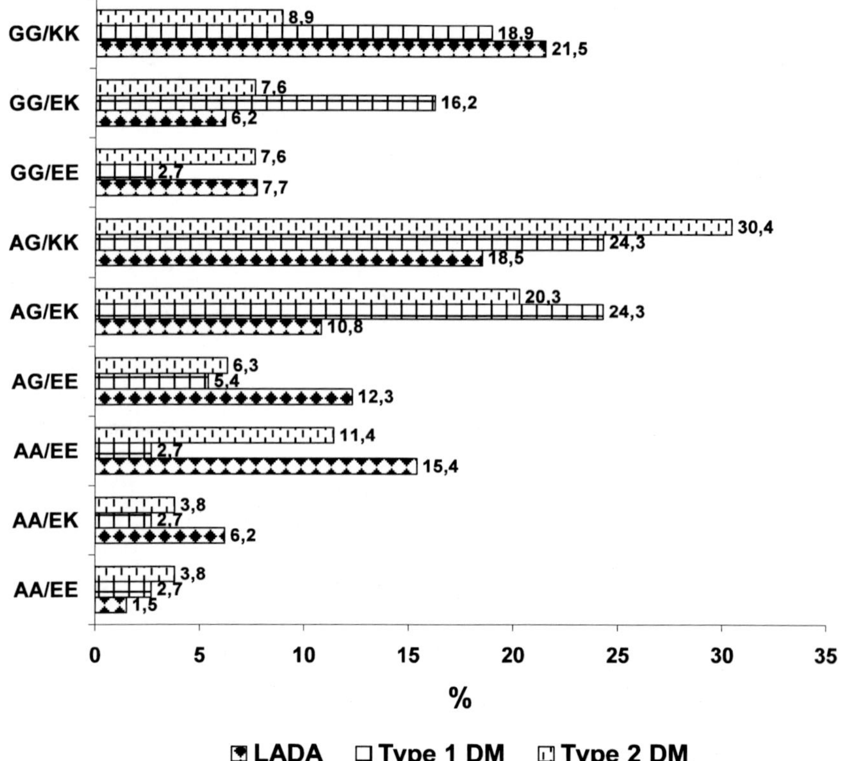

Figure 5. Distribution of 49A/G polymorphisms of *CTLA4* gene and E23K of *KCNJ11* gene in patients with different clinical variants of DM course.

Distribution of patients by carriage of polymorphisms of C1858T of *PTPN22* gene and E23K of *KCNJ11* gene is presented on Figure 6.

The significant distinctions in genotypes distribution with C1858T of *PTPN22* gene and E23K of *KCNJ11* gene ($\chi^2 = 11.146$; df = 16; p = 0.025) among the compared groups of patients were shown. The pair-wise comparison of DM clinical variants shown the significant distinctions between the patients with LADA and type 2 DM ($\chi^2 = 5.233$; df = 8; p = 0.264). The significant distinctions in this polymorphisms between the patients with LADA and type 1 DM, also between the patients with type 1 and 2 DM were not found ($\chi^2 = 5.928$; df = 8; p = 0.205; $\chi^2 = 6.303$; df = 8; p = 0.178, accordingly).

Distribution of patients by carriage of polymorphisms of 49A/G of *CTLA4* gene, C1858T of *PTPN22* gene and E23K of *KCNJ11* gene is presented on Figure 7.

The significant distinctions in distribution of the genotypes with C1858T of *PTPN22* gene and 49A/G of *CTLA4* gene, and E23K of *KCNJ11* gene (χ^2 = 56.923; df = 46; p = 0.000) among the compared groups of patients were shown. The pair-wise comparison of DM clinical variants shown the significant distinctions between all investigated groups: the patients with LADA and type 1 DM (χ^2 = 23.978; df = 23; p = 0.000); the patients with LADA and type 2 DM (χ^2 = 19.341; df = 23; p = 0.000); the patients with type 1 and type 2 DM (χ^2 = 21.071; df = 20; p = 0.000); that indicates the genetic independence of all these clinical DM forms.

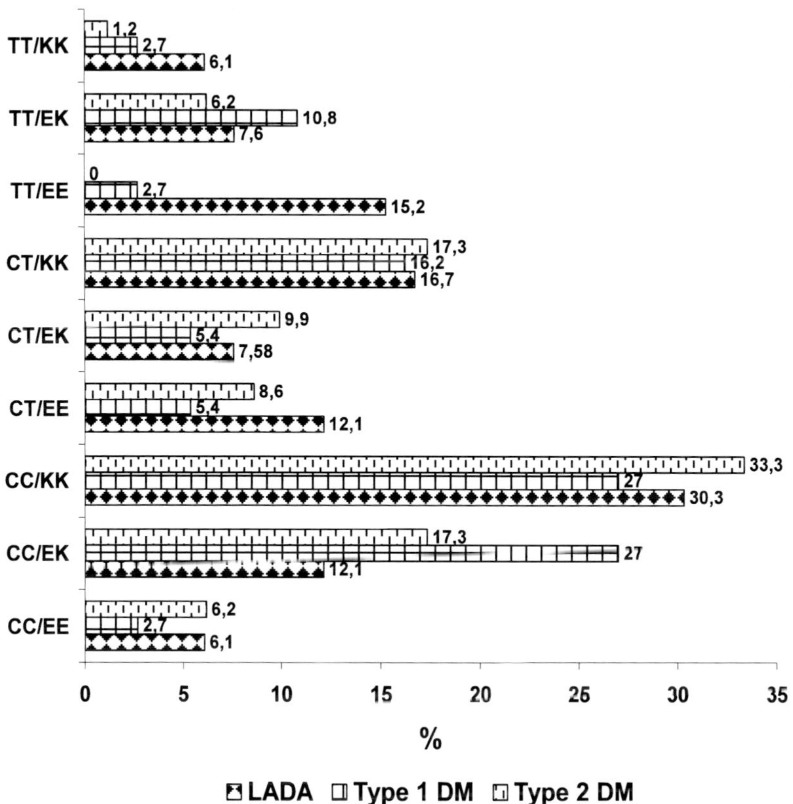

Figure 6. Distribution of polymorphisms of C1858T of *PTPN22* gene and E23K of *KCNJ11* gene in patients with different clinical variants of DM course.

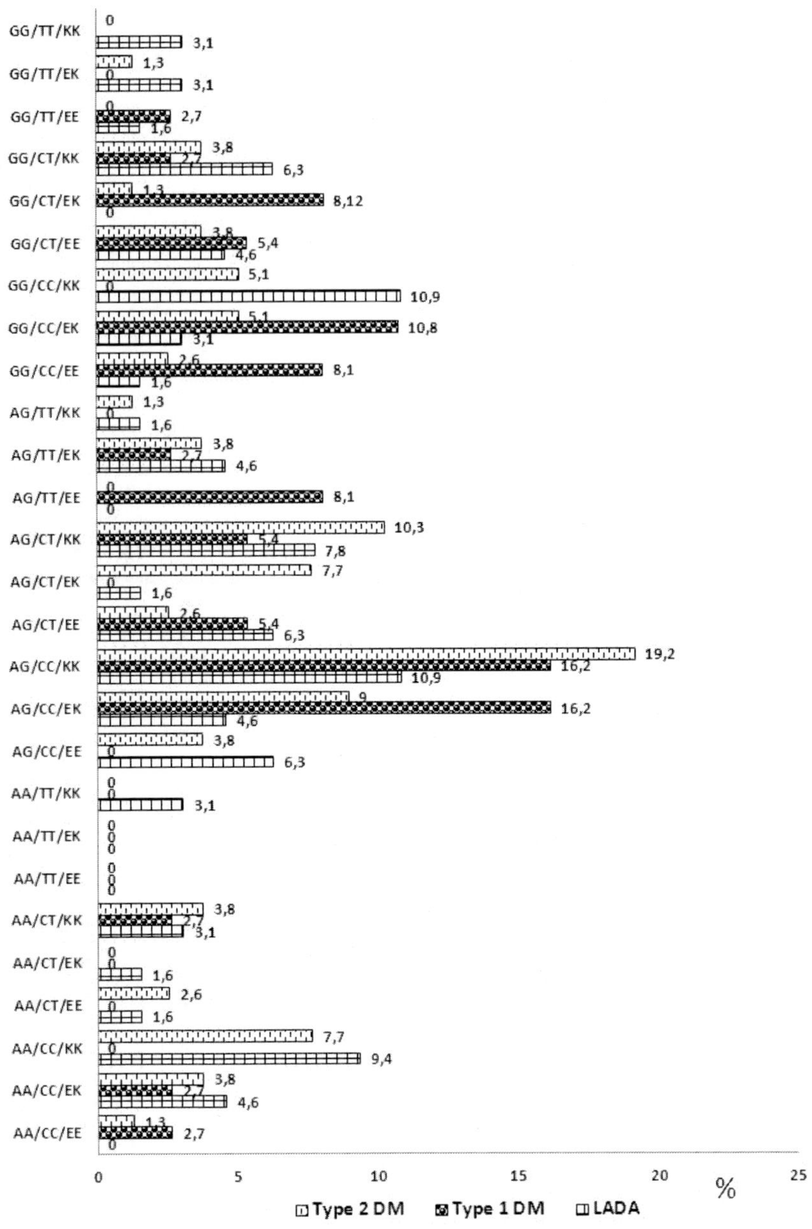

Figure 7. Distribution of polymorphisms of 49A/G of *CTLA4* gene, C1858T of *PTPN22* gene and E23K of *KCNJ11* gene in patients with different clinical variants of DM course.

However, if the genetic independence of type 1 and 2 DM is well-proven, the genes of susceptibility to these disease types are selected, a question about LADA genetic determination and reasons for its occurrence remain open to date. It is assumed for today, that LADA is the variant of type 1 DM course [81]. The molecular-genetic research, conducted by us, shown LADA genetic independence, confirming the results of genetic [51] analysis, obtained before.

Absence of significant distinctions among the investigated single gene polymorphisms in the patients with type 1 and 2 DM completely corresponds to the genetic analysis results, which indicated the presence of 59% common genes in type 1 and 2 DM determination [41]. Taking into account the fact that type 1 and 2 DM are polygenetic diseases, not only the genes, which caused one or another DM type development, the genes combination is significant too. Selection tension against type 1 DM reduction, caused by the insulin therapy introduction, and positive direction of genes of susceptibility to type 2 DM selection resulted not only in increase of the genes of susceptibility to type 1 and 2 DM in population, but also in their combination variants increase. It leads, in particular, to growth of such genetically independent DM form as LADA and increases its prevalence among the DM patients.

Thus, the conducted research showed the selection role in changing of prevalence of type 1 and 2 DM in population and in evolution and in origin of DM clinical forms.

DISCUSSION

In this work the peculiarities of different DM types selection, which appear in different ontogenetic periods and differ in morbidity and genetic determination, were studied. Thus, if type 1 DM, without an adequate insulin therapy, results in fatal outcome and, to some extent, can be partly associated with the factors, regulating the population amount, type 2 DM, vice versa, promoted the increase of survival rate of carriers of this trait in the past historical periods with food deficiency. According to J.V. Neel conception, the special genes (thrifty gene), which allowed a man to use effectively the limited food resources, results in obesity and type 2 DM development in the conditions of food abundance [28].

The conducted research showed, that selection in favour of diseases, which commonly start in a post-reproductive period, to which type 2 DM belongs, exists till the present time. The reasons, determining the positive direction of type 2 DM selection, remain unknown till today and, probably, can be of interest for further researches. The selection positive direction was accompanied by significant growth of their prevalence in population over a generation.

Peculiarities of type 1 DM selection were influenced by the modern medicine, which succeeded in compensating the disease with the insulin therapy. This enabled a patient to survive and produce geniture. Type 1 DM in the "pre-insulin period" belonged to the sub-lethal traits. Its manifestation usually occurs in the pre-pubertal or pubertal period, and then during half a year – a year such patients died from the hyperglycemic come. The insulin therapy allowed to correct the disease course significantly, however, the negative consequences cannot be completely neutralized till today, that changed type 1 DM from sub-lethal trait into the trait of reduced adaptability. The peculiarities of clinical course of type 1 DM result in lower geniture survival rate that determines the low relative adaptability of disease and, accordingly, the negative selection direction.

The insulin therapy introduction in 50th years of 20th century into the common health care practice has led to decrease of selection tension against type 1 DM and increase of the population prevalence of this type of disease. Growth of type 1 DM prevalence proceeded up to 1995, approximately for two generations (fifty years). After 1995 the rapid growth of type 1 DM prevalence has stopped, apparently, the existent selection against type 1 DM tension played a certain role in it. Consequence of the ongoing processes is not only increased prevalence of type 1 DM in population, but also the increased clinical DM heterogeneity. Increase in frequency of genes of susceptibility to type 1 DM in population, caused by selection tension reduction, was not only the basis for increase of population prevalence of this disease type, but also promoted probability of genes occurrence in proband, determining development of both DM types. The result of this phenomenon could be a formation and growth of prevalence in DM structure of such clinical form of disease as LADA [51], which heritability is described by the polygene threshold model parameters. In its inheritance the essential role belongs to the genetic factors (59.2%), there are nonlinear genetic intraloci and interallelic (GD = 1.2%) interactions and influence of number of genes with evident effect in determination of this form of disease is possible. Although LADA is an independent DM genetic form, but, according to Ch. Smith model testing, there is approximately the same amount of common genes of susceptibility to type 1 and 2 DM (65.3 and 66.1%, accordingly) in its genetic control system.

Conducted researches of polymorphisms of C1858T of *PTPN22* gene and 49A/G of *CTLA4* gene, associated with type 1 DM in patients with different DM clinical forms, show that those polymorphisms are associated with type 2 DM and with LADA. Analogical data on associations of this polymorphisms with type 1 and 2 DM and LADA were obtained also in other populations in different countries of the world [67-72]. It should be noted, that the insulin therapy introduction into the common health care practice for correction of type 1 DM course was observed in all countries of the earth. And such disease compensation allowed to reduce considerably the selection tension against this type of DM in all countries. It increased the frequencies of genes of susceptibility to type 1 DM in all populations. Therefore, the association of polymorphism of C1858T of *PTPN22* gene and type 1 DM is not local, but observed in different countries. The model, presented before, of type 1 DM inheritance [41] indicated the linked factors, determining the hereditary susceptibility to this type of disease (high values of epistatic components). It is presently known, that antigens of the HLA system are the major contributors to type 1 DM development [42]. The HLA system antigens are codominantly inherited and are linked factors [82]. Other genes of susceptibility to type 1 DM, such as *CTLA4, PTPN22, IL2RA, IFIH1, IL2, PTPN2* and *KIAA0350*, are localized on different chromosomes [42]. Taking into account that these genes are linked considerably less, than antigens of the HLA system, it is logically to assume that *CTLA4, PTPN22, IL2RA, IFIH1, IL2, PTPN2* and *KIAA035* genes can form different combinations with the genes of susceptibility to type 2 DM in a proband. From this point of view, it is understandable that LADA patients more frequently than the ones with type 1 DM have such gene polymorphisms as C1858T *PTPN22* and 49A/G *CTLA4* genes. The observed phenomenon corresponds to ideas of numerous authors, that LADA has an autoimmune nature, but differs from type 1 DM by type of insulin dependence and by distribution of susceptibility to DM alleles [49]. Frequency of C1858T of *PTPN22* gene and 49A/G of *CTLA4* gene polymorphisms in type 2 DM is considerably less than in LADA and is analogical to the one in type 1 disease, although carriage of this polymorphisms is not significant in pathogenesis of type 2 DM. It can be explained by selection tension reduction against type 1 DM, caused by the insulin therapy implementation, accompanied by increase of population genes frequency, causing the type 1 DM development. It can logically explain the fact of association of C1858T of *PTPN22* and 49A/G of *CTLA4* genes polymorphisms with type 2 disease, because the selection tension reduction promoted the occurrence probability of the genes of susceptibility to type 1 DM in patients with type 2 DM. However, the presence of one or another genes polymorphism, determining the immune response disorders in patients with type 2 DM, does not cause any autoimmune aggression against the pancreatic islet cells, as the latter is caused by a combined effect of multiple mutations, causing immune response disorder.

Because the genetic control system LADA, in according the heredity model, has a bigger part of common genes with type 2 DM, it was logical to study an association of LADA, type 1 and 2 DM with polymorphism of E23K of *KCNJ11* gene, which determines the development of insulin insufficiency in a proband which is a gene, taking part in the susceptibility to type 2 DM forming [42]. The obtained results were analogical to our data about the association of polymorphisms of C1858T of *PTPN22* gene and 49A/G *CTLA4* gene and DM. The significant association of this mutation with all clinical variants of DM course (Table 10) was also shown. Thus, the conducted molecular-genetic researches confirmed positions of LADA and type 1 and 2 DM inheritance models about existence of common genes in their genetic control systems.

The distribution of the patients with different clinical variants of DM course by the polymorphisms of C1858T of *PTPN22*, 49A/G of *CTLA4* and E23K of *KCNJ11* genes shown the significant distinctions ($\chi^2 = 56.923$; df = 46; p = 0.000) in the distribution of patients with different clinical variants of disease course by the genotypes, which were investigated. The pair-wise comparison of patients groups revealed the significance of distribution of patients by the carriage of polymorphisms, which were studied among all investigated groups, that proves the genetic independence of all these DM forms. Conducted molecular-genetic research showed LADA genetic independence, confirming our genetic analysis results [51]. In addition it should be noted, that larger, than among the patients with type 1 and 2 DM, number of homozygote carriers of all three mutant alleles is characteristic for LADA. The obtained data confirm Hamaguchi assumption, that LADA has an autoimmune nature, but differs from type 1 DM by types of the insulin dependence and distribution of alleles, susceptible to DM [49]. And exactly the high frequency of mutant homozygote carriers of immune response genes among the patients with LADA, and only partial availability of the HLA genes system in them, which is forming the susceptibility to type 1 DM, determine late development of this DM form and its less severe course. The high percentage of homozygotic carriers of mutations, forming susceptibility to type 1 and 2 DM, is a bright confirmation of assumption about the selection influence on evolution of DM clinical forms. Exactly the high percent of homozygotic carriage of polymorphic alleles, simultaneously by all three investigated positions in LADA patients, illustrates the position that this form of DM occurs as a result of selection action because of frequencies increase of the genes of susceptibility to both type 1 and 2 DM in population.

CONCLUSION

Thus, natural selection has defined the prevalence dynamics and influences formation of the polygenic diseases heterogeneity, to which DM belongs.

REFERENCES

[1]   F. Vogel, A. G. Motulsky, *Human genetics. Problems and Approaches,* Springer-Verlard, Berlin Heindelberg, New York, Tokio (1986).

[2]   F. J. Ayala, J. A. Kiger, *Modern genetics,* University of California, Davis, Second Edition (1984).

[3]   Y. P. Altukhov, Y. G. Rychkov, *Journal of common biology,* 31, 507 (1970).

[4]   Shmal'gauzen, *Cybernetic questions of biology.* Science, Novosibirsk (1968).

[5]   Y. P. Altukhov, *Bulletin of genetics and selectors society,* 8, 40 (2004).

[6]   O. L. Kurbatova, *Materials of the third anthropological reading to the 75 year from the day of birth of academician V. P. Alekseeva «Ecology and demography of man in the past and present»* in press (2004) "Demographic genetics of megapopulation: changeability of selection parameters," The third anthropological reading to the 75year from the day of birth of academician V. P. Alekseeva «Ecology and demography of man in the past and present» 15-17, November. 2004 (Moscow, Russia). pp. 259-262.

[7]   WF Bodmer and LL Cavalli-Sforza, *Genetics, Evolution, and Man,* WH Freeman and Company, San Francisco (1976).

[8]   N. Kucher, O. L. Kurbatova, *Genetics,* 22, 304 (1986).

[9]   L. A. Atramentova, L. I. Fedchun and S. A. Povolockiy, *Genetics,* 29, 520 (1993).

[10] J. F. Crow, *Human Biol.* 30, 1 (1958).

[11] N. P. Dubinin, I.I. Karpec and V.N. Kudryavcev, *Genetics, behaviour, responsibility*, Politizdat, Moscow (1982).

[12] Y. P. Altukhov, *Genetic processes in populations*, IKC "Akademkniga," Moscow (2003).

[13] L. S. Penrose, *Proc. Roy. Soc. Ser. B.,* 159, 93 (1963).

[14] V. M. Kornac'kiy *Problems of health of society and the life duration*, Kiev (2006).

[15] O. L. Kurbatova, *Theses of lectures of First (third) Russian congress of medical geneticists* in press (1994). "Some tendencies of transformation of adaptive genetic-gemographic structure of multinational populations of Russia and CIS." The First (third) Russian congress of medical geneticists, 14-16, Dec. 1994, (Moscow, Russia) p.112.

[16] Y. E. Dubrova, T. M. Karafet, R. I. Sukernik and T. V. Gol'tsova, *Genetics*, 26, 122 (1990).

[17] L. I. Bol'shakova and A. A. Revazov, *Genetics*, 24, 340 (1988).

[18] O. L. Kurbatova and E. Y. Pobedonosceva, Genetic-demographic processes at urbanization: migration, outbreeding and marriage assortativeness. In: Y. P. Altukhov *Heredity of man and environment.* Science, Moscow, 2, pp.7–22 (1992).

[19] R. L. Berg and S. N. Davidenkov, *Heredity and inherited diseases of man*, Leningrad (1971).

[20] V. Surin, *News*, № 224, (2003).

[21] On the threshold of XXI century: Report about the world development of 1999/2000. 23.

[22] Y. Vnutskikh, *New ideas in philosophy,* 13, 2004 at http://www.psu.ru/psu/files/1641/09_Vnutskih.doc.

[23] V. S. Baranov, *Soros educational magazine.* № 3, 3 (1999).

[24] E. Mayr, *The growth of biological thought: diversity, evolution; and inheritance,* Cambridge: Harvard University Press (1985).

[25] G. W. Williams and R. M. Nesse, *Q. Rev. Biol.*, 66, 1 (1991).

[26] R. M. Nesse and G. C. Williams, *Why we get sick? The new science of Darwinian medicine*, Vintage Books, New York (1996).

[27] R. M. Nesse, *World Psychiatry*, 7 (2002).

[28] J. V. Neel, *Am. J. Hum. Genet.*, 14, 4, 353 (1962).

[29] C. Liberman, *Russian cardiologic journal*, 35 (2002).

[30] M. L. Harris, W. C. Hadden, W. C. Knowler and P. H. Bennett, *Diabetes*, 523 (1987).

[31] J.-M. Ekoé, P. Zimmet, and D. R. R. Williams, *The epidemiology of diabetes mellitus: an international perspective*, Wiley and Sons, Chichester, UK (2001).

[32] Y. P. Altukhov, *Man*, № 6, 2003 at http://vivovoco.rsl.ru/VV/PAPERS/ MEN/GENETICS.HTM.

[33] V. S. Baranov and E. V. Baranova, *Ecological genetics*, 22, (2004).

[34] E. K. Ginter, A. A. Revazov and A. N. Petrin, *Influence of urbanization on the load of the inherited diseases in populations*. In press Y. P. Altukhov "Heredity of man and environment." Moscow: Science, 2, 22, (1992).

[35] V. N. Serov, G. M. Burduli and O. G. Frolova, *The reproductive losses*. Moscow: Success (1997).

[36] N. Zhilka, T. Irkina and V. Teshenko, *The condition of reproductive health in Ukraine (medico-demographic survey)*. Kiev (2001).

[37] N. B. Kerimi, A. S. Sergeev, A. G. Mazovetskii and T. L. Kuraeva, *Genetica*, 20, 166, (1984).

[38] S. Ghosh and N. J. Schork, *Diabetes*, 45, 1, (1996).

[39] M. Mc Carthy, P. Froguel and G. A. Hitman, *Diabetologia*, 37, 959, (1994).

[40] S. Shtandel, S. Finogenova and L. Atramentova, *Diabetologia*, 41, Suppl. 1, A104, (1998).

[41] S. A. Shtandel, L. A. Atramentova, S. A. Finogenova and A. R. Gevorkyan, *Cytology and genetics*, 34, 34, (2000).

[42] T. M. Frayling and M. I. McCarthy, *Diabetologia*, 50, 2229, (2007).

[43] M. I. Balabolkin, E. M. Klebanova and V. M. Kreminskaya *Treatment of diabetes mellitus and his complications: Guidance for doctors*, "Medicine," Moscow, (2005).

[44] P. Zimmet and D. McCarty, *IDF Bulletin*, 39, (1995).

[45] U.K. Prospective Diabetes Study (UKPDS) Group. Intensive blood-glucose control with sulphonylureas or insulin compared with conventional treatment and risk of complications in patients with type 2 diabetes (UKPDS 33), *Lancet*, 352, 837 (1998).

[46] M. Mkrtum'yan, *The attending physician*, № 10, 32, (2003).

[47] S. A. Shtandel, *Natural selection of metabolic syndrome basic clinical components in modern terms*. (2012). Ch. M. Lopez Garsia and P. A. Perez Gonzalez, "Handbook on Metabolic Syndrome, Classification, Risk Factors and Health Impact." New York: Nova Biomedical. 333, (2012).

[48] Report of Expert Committee on the Diagnosis and Classification of Diabetes Mellitus. *Diabetes Care*, 25, Suppl. 1, 5, (2002).

[49] K. Hamaguchi, Y. Kusuda, N. Abe and T. Sakata, *Chin. J. Pathphysiol*, 17, 768, (2001).

[50] K. J. Basile, V. C. Guy, S. Schwartz and S. F. Grant, *Curr Diab Rep.*, 14, 550, (2014).

[51] S. A. Shtandel' and T. M. Tikhonova, *Cytology and Genetics*, 42, 358, (2008).

[52] N. D. Tron'ko and A. D. Chernobrov, The basic indexes of activity of endocrinologic service of Ukraine for 2007. Kiev, 31 tabl, (2008).

[53] N. D. Tron'ko and A. D. Chernobrov, *The basic indexes of activity of endocrinologic service of Ukraine for 2000.* Kiev, 30 tabl, (2001).

[54] N. D. Tron'ko and A. D. Chernobrov, *The basic indexes of activity of endocrinologic service of Ukraine for 2001.* Kiev, 30 tabl, (2002).

[55] N. D. Tron'ko and A. D. Chernobrov, *The basic indexes of activity of endocrinologic service of Ukraine for 2002.* Kiev, 30 tabl, (2003).

[56] N. D. Tron'ko and A. D. Chernobrov, *The basic indexes of activity of endocrinologic service of Ukraine for 2003.* Kiev, 30 tabl, (2004).

[57] N. D. Tron'ko and A. D. Chernobrov, *The basic indexes of activity of endocrinologic service of Ukraine for 2013.* Kiev, 31 tabl, (2014).

[58] The basic indexes of the specialised endocrinologic help to the population of the Ukrainian Soviet Socialist Republic in 1989 – 1990. The Kiev Scientific research institute of endocrinology and metabolism, Kiev, 30 tabl, (1991).

[59] Y. Boyarskiy Population and methods of his study. *Statistica,* Moscow (1975).

[60] W. Zheng, *Diabetes*, 54, 906, (2005).

[61] H. Donner, H. Rau, P.G. Walfish, T. Siegmund, R. Finke, J. Herwig, K. H. Usadel and K. Badenhoop, *Journal of Clinical Endocrinology and Metabolism*, 82, 143, (1997).

[62] Ch. Ezenwaka, R. Kalloo, M. Uhlig, R. Schwenk and J. Eckel, *Journal of Endocrinology*, 185, 439, (2005).

[63] M. Bland and D. G. Altman, *BMJ*, 320, 1468, (2000).

[64] N. G. Goyda, *Scientific basis of the development of the medicosanitary help system to women with an extragenital pathology.* Avtoessay of DMed thesis: 14.02.03. National medical university by name of O.O. Bogomol'tsa. Kiev, (2000).

[65] E. D. Cherstvoy, G. I. Kravcova and G. I. Lazyuk, *Diseases of fruit, new-born and child.* Higher school, Minsk (1991).

[66] N. Bottini, L. Musumeci, A. Alonso, S. Rahmouni, K. Nika, M. Rostamkhani, J. MacMurray, G. F. Meloni, P. Lucarelli, M. Pellecchia, G. S. Eisenbarth, D. Comings and T. Mustelin, *Nat Genet.*, 36, 337, (2004).

[67] M. Fedetz, F. Matesanz, A. Caro-Maldonado, I. I. Smirnov, V. N. Chvorostinka, T. A. Moiseenko and A. Alcina, *Tissue Antigens*, 67, 430, (2006).

[68] Petrone, C. Suraci, M. Capizzi, A. Giaccari, E. Bosi, C. Tiberti, E. Cossu, P. Pozzilli, A. Falorni, R. Buzzetti and NIRAD Study Group, *Diabetes Care*, 31, 534, (2008).

[69] Cervin, V. Lyssenko, E. Bakhtadze, E. Lindholm, P. Nilsson, T. Tuomi, C. M. Cilio and Leif Groop, *Diabetes*, 57, 1433, (2008).

[70] D. Abramov, I. I. Dedov, D. Y. Trophimov, M. N. Boldyreva, T. L. Kuraeva and L. P. Alekseeva, *Diabetes Mellitus*, № 3: (2007), http:// dmjournal.ru/_mod_files/_upload/SD2007_3_2.pdf.

[71] M. Krokowski, J. Bodalski, A. Bratek, P. Machejko and S. Caillat Zucman, *Diabetes Metab.*, 24, 241, (1998).

[72] D. A. Chistiakov, K. V. Savost'anov and V. V. Nosikov, *BMC Genet.*, 2, (2001), http://www.biomedcentral.com/1471-2156/2/6.

[73] C. Guja, S. Marshall, K. Welsh, M. Merriman, A. Smith, J. A. Todd and C. Ionescu Tirgoviste, *J. Cell Mol. Med.*, 6, 75, (2002).

[74] Fajardy, A. Vambergue, C. Stuckens, J. Weill, P. M. Danze and P. Fontaine, *Eur. J. Immunogenet.*, 29, 251, (2002).

[75] M. Mochizuki, S. Amemiya, K Kobayashi, Y. Shimura, T. Ishihara, Y. Nakagomi, K. Onigata, S. Tamai, A. Kasuga and S. Nanazawa, *Diabetes Care*, 26, 843, (2003).

[76] F. K. Kavvoura and J. P. Ioannidis, *Am. J. Epidemiol.*, 162, 3, (2005).

[77] T. Yanagawa, T. Maruyama, K. Gomi, M. Taniyama, A. Kasuga, Y. Ozawa, M. Terauchi, H. Hirose, H. Maruyama and T. Saruta, *Autoimmunity*, 29, 53, (1999).

[78] O. Cinek, P. Drevnek, Z. Sumnk, B. Bendlová and S. Kolousková, *Eur. J. Immunogenet.*, 29, 219, (2002).

[79] Dedov, L. I. Kolesnikova and T. P. Bardimova, *Polymorphism of CTLA4 (49 A/G) gene in patients of buryat population with the 1 type diabetes Materials* in press (2006). The 13th International Congress on circumpolar health, June 12-16 2006. (Novosibirsk) p. 63

[80] Y. Y. Li, *Mol. Biol. Rep.*, 141, (2013).

[81] Dedov and M. V. Shestakova, *Diabetes mellitus: diagnostics, treatment, prophylactic.* Moscow (2011).

[82] V. V. Nosikov and Y. A. Seregin, *Molecular biology*, 42, 867, (2008).

In: Natural Selection and Genetic Drift   ISBN: 978-1-63484-331-7
Editor: Joshua Richardson   © 2016 Nova Science Publishers, Inc.

Chapter 3

GENETIC DRIFT AMONG NATIVE PEOPLE FROM SOUTH AMERICAN GRAN CHACO REGION AFFECTS INTERLEUKIN 1 RECEPTOR ANTAGONIST VARIATION

Cecilia Inés Catanesi and Laura Angela Glesmann*
Laboratory of Genetic Diversity, IMBICE, La Plata, Argentine

ABSTRACT

Genetic variation is generally responsible for ethnic differences in certain diseases, including inflammatory processes. The antagonist of cytokine IL-1, IL-1Ra, has been widely studied among Caucasian and African populations for genetic polymorphisms, and interethnic differences have been documented. However, the variation and genotype distribution of polymorphisms from these genes among South American Amerindians are thus far unknown. We present the results for a VNTR located in the IL-1Ra second intron, in a sample of 169 individuals belonging to 5 Native American populations from Argentina and Paraguay, identified as native according to their self designation, and their geographic location. We also compare this data with the results obtained from a sample of non-native Argentinian people. Among the five known alleles of the VNTR, we found only two (alleles 1 and 2) in the native populations from Gran Chaco, and heterozygosity was 19%.

---

* Email: ccatanesi@imbice.gov.ar.

The allele 2 which is considered proinflammatory (IL-1Ra * 2) has been found in homozygosity at a considerable frequency among native individuals. However, the association of this allele with inflammatory disease previously demonstrated for other populations of the world, might not be acting in the same way for native people, probably due to local adaptation. This would indicate that the allele 2 will probably not have a negative influence on individuals of native origin who have homozygous genotype 2-2. On the contrary, few records on inflammatory disease are available for the native people.

It seems that the increment on allele 2 is not related to any adaptive process but to genetic drift, that changes randomly the allele frequencies of different genetic regions along the genome. The effect of genetic drift has already been demonstrated with genetic markers located in autosomes, X and Y chromosomes. These results indicate that we must be very cautious when studying populations that passed a process of genetic drift, which can become a confounding factor in epidemiological studies. This information will contribute to a future understanding of the association of this polymorphism with disease, and its incidence in different ethnic groups.

TEXT

The native human populations inhabiting the American territory have arrived to the continent through different migratory events, and have been involved in social and cultural interactions with other human groups. This exchange can be partially reflected in their current genetic structure. However, other processes than genetic flow have played an important role in the microevolutive change of these populations, and genetic drift might be, even nowadays, one of the most important processes acting on them.

Before the arrival of the European colonizers, the Gran Chaco region possessed a rich cultural and linguistic diversity. But the initial contact between Native Americans and Europeans five hundred years ago, initiated a dramatic reduction of the native populations, as happened in other parts of the American continent (Martínez-Sarasola 2004). Many groups were completely wiped out, while others have introduced some genetic admixture from non-native groups (Mulligan et al. 2004) giving rise through the centuries to the current Latin American populations, which share three main ancestral origins: Native American, European, and African (Salzano and Bortolini 2002; Rondón et al. 2008). In this chapter we present the analysis of some extant native groups inhabiting the South American Gran Chaco region, and the genetic drift process that can be found through genetic analysis. Information on these populations is not always concordant with the conclusions on their origin and history (Torroni et al. 1993; Horai et al. 1993; Zago et al. 1995), in part due to the genetic drift process, which generates a strong differentiation among tribes.

Figure 1. Geographic location of the phytogeographic Gran Chaco region.

The Gran Chaco is a plain, very wide phytogeographic region of approximately 100.000 km^2, which includes part of Bolivia, Paraguay, and northeast of Argentina (Figure 1). About 5.000 years ago it was colonized by several nomadic native groups of hunting, fishing and gathering habits: Abipon, Mocoví, Qom, Mbya, and Pilagá belonging to Guaycurú linguistic group, Wichí, Chorote, and Chulupí of the Mataguayo linguistic family, and the Lule-Vilela group in the western part, corresponding to current Argentinian territory. In the northeast, the Ayoreo and Lengua, live in the region that belongs to Paraguay. These ethnic groups did not practice written language, therefore the historical record begins with the arrival of Spanish colonizers, and before that moment, information can be obtained only from oral transmitted tales (Martínez-Sarasola 2004; Tissera 2008). These native groups show an important diversity of spoken languages as a result of complex interactions and interchanges (Jurado Medina et al. 2014).

BACKGROUND

Molecular biology offers a wide variety of coding and non coding (neutral) DNA markers for analyzing the diversity among individuals of a population, and among populations of a region. There is a background of information on DNA markers for Gran Chaco populations including several SNPs (single nucleotide polymorphisms), STRs (short tandem repeats, also called microsatellites), and insertions-deletions. SNPs are small changes in the DNA sequence affecting only one nucleotide. Usually, SNP mutation rates are moderate to low. STRs are short sequences tandemly repeated, they are highly variable and highly informative for studying genetic diversity and evolutionary processes. Insertion-deletion markers are sequences that are either present or absent, and can extend from only one or few nucleotides to several hundreds of them. An important amount of information has been obtained from non coding markers in Y chromosome, mitochondrial DNA, autosomes and X chromosome, while much fewer information is available from coding regions, such as blood antigens and HLA genes (Goicoechea et al. 2001; Dejean et al. 2004).

UNIPARENTAL Y CHROMOSOME VARIABILITY

Male specific region of Y chromosome has been widely studied throughout the world, allowing to disentangle the genetic structure of different populations. The diversity of Gran Chaco native people has been analyzed through uniparentally inherited markers localized in Y chromosome, SNP and STR polymorphisms.

For the native populations of Gran Chaco, the search of autochtonous Q1a3a haplogroup has been firstly focused among different Chaco groups to determine either a native or a non-native origin of this chromosome. This haplogroup is determined by M3 SNP marker. Among those considered native, a set of STR markers has been usually genotyped to determine the diversity of each analyzed group.

An analysis on three different ethnic groups, Pilaga, Wichí, and Toba on SNP and STR markers showed genetic drift as a powerful evolutionary force for these seasonal hunters living in small bands (Demarchi and Mitchell 2004). The Native American-specific M3 marker was carried by 76.9% of the individuals, with a moderately high intergroup variation based on Y-chromosome STRs (10.7%).

In this way, two populations of Wichí origin living in Formosa province, 70 kilometers away from each other, were analyzed by Ramallo et al. (2009), showing a frequency of the Q1a3a haplogroup of 72,7% and 81,6%, and a rich lineage variability regionally distributed. Allelic variation was also non coincident. Their nomadic way of life and their habit to live in small groups has been clearly a strong force driving to genetic drift. Moreover, the Wichí people show a distribution of partialities called "Wichí ethnic complex" with certain linguistic differences (Braunstein 2006; Ramallo et al. 2009).

However, a small sample belonging to another ethnic group, did not show such influence. The Mocoví people living south to Gran Chaco, in Santa Fe province, were also analyzed for Y markers including two SNPs (M3 and M346) and ten STRs (Glesmann et al. 2011). The M3-T transition was present in 52% of the individuals, and STR haplotype diversity was 99.69%. In this case, the Mocoví Y-chromosomes still retain an interesting variability, with some of the M3-T haplotypes not found in other Amerindian groups. This considerable amount of haplotype variability is likely to be originary from this population.

X CHROMOSOME VARIABILITY

Due to its particular mode of inheritance, and its lower recombination rate compared to autosomes, the patterns of genetic variation of different types of markers specifically located in the X chromosome can result highly informative for population studies. Indeed, its special characteristics give a chance for analyzing genetic variability from a different point of view (Bourgeois et al. 2009; Ribeiro Rodrigues et al. 2009, 2011).

A study on X chromosome variation was performed including five Chaco ethnic groups: from Argentina, Wichí and Chorote from Salta province, and Mocoví from Santa Fe; from Paraguay: Lengua and Ayoreo (Catanesi et al. 2007). This analysis showed significant differences among these populations, with high Fst and Rst values, probably due to the drift process. The differentiation was related to the geographic location of populations, grouping those from Salta together, and those from Paraguay together, with Mocoví resulting separated from the rest (Catanesi et al. 2007).

A more recent study included another Wichí population living in a region called "Impenetrable" due to its hard climatic conditions and dense vegetation, in the Chaco province, and a Mocoví population (the same that was analyzed for Y chromosome) living in the south of Gran Chaco (Glesmann et al. 2013). A high proportion of homozygotes and a marked linkage disequilibrium was found especially in Wichí, with differences in modal alleles and frequencies between both populations. On the other hand, a higher proportion of variation was observed in Mocoví. It has been reported that the Mocoví people belonging to the studied population are currently taking part of the neighbor non-native society through work and education. This social integration might be responsible for a cultural change among Mocoví (Franceschi and Dasso 2010). Although the genetic flow between Mocoví and non-natives might be occurring (Citro 2006), a more important process of genetic drift may be reflected in the X chromosome variation reported, especially among Wichí. The Wichí people from Chaco province not only live isolated from other native and non-native groups, but also display an irregular distribution in small bands along the territory. Their geographic isolation and the extreme environmental conditions may be considered as the major factors contributing to the population differentiation (Glesmann et al. 2013).

UNIPARENTAL MITOCHONDRIAL VARIABILITY

Native Americans share five different maternally inherited mitochondrial DNA haplogroups: A, B, C, D, and X (Schurr et al. 1990; Torroni et al. 1993; Santos et al. 1996), which are present in different proportions depending on the demographic background of each population (Bailliet et al. 2004; Avena et al. 2012; Wang et al., 2008; Pauro et al. 2010; Yang et al. 2010; Motti et al. 2013). An unbalanced proportion of male and female native uniparental lineages has been clearly described, as a consequence of mixed marriages of European males with native women through the last 5 centuries, thus making maternal haplogroups of native origin to widespread in different regions (García and Demarchi, 2006; 2009, Pauro et al. 2010; Yang et al. 2010).

Interestingly, a loss of mitochondrial variability has been described for certain Gran Chaco communities, including Wichí from Chaco province and Ayoreo from Paraguay. The random action of genetic drift or a bottleneck effect has been proposed as responsible for this reduction (Demarchi et al. 2001; Demarchi and Mitchell 2004).

AUTOSOME NON CODING VARIATION

Studying different genomic compartments has contributed to the understanding of the evolutionary processes occurring among South American Gran Chaco native populations. Different autosomal STR markers have shown in general a drop in the number of alleles and, consequently, an excess in the proportion of homozygote individuals. Variation in modal alleles for each native population, and a drastic reduction in allele number were found, particularly among Wichí (Tourret et al. 2000; Catanesi et al. 2001). As a consequence of genetic drift, a relatively poor correlation with geographic location of the tribes was more notable when analyzing autosomal variation (Zago et al. 1996; Tourret et al. 2000; Catanesi et al. 2001) than X chromosome variation (Catanesi et al. 2007).

AUTOSOME CODING VARIATION IN THE INTERLEUKINE 1 RECEPTOR ANTAGONIST

Coding genetic variation is generally responsible for ethnic differences in certain diseases. Natural selection changes allele frequencies according to environmental conditions, thus modifying the diversity of a population under selection (Hedrick 2000). Information on coding regions is scarce for native populations from Gran Chaco region, but the HLA region has been reported to present a low variability of dinucletide STRs for three Chaco tribes: Wichí, Chorote, and Toba (Dejean et al. 2004), and metabolic genes have also been studied (Bailliet et al. 2007).

The inflammatory processes can be more prevalent among individuals who belong to a certain ethnic group, thus needing a specific medical treatment. Interleukin-1 (IL-1) is a cytokine secreted by macrophages and other cell types, playing an important role in the inflammatory response of virtually all cells and organs, as a major pathogenic mediator inducing pain (Dinarello et al. 2012; Gabay et al. 2010; Sims and Smith, 2010; Garlanda et al. 2013). The IL-1 gene family comprises several genes including three closely related IL-1A, IL-1B, and IL-1Ra encoding respectively two proinflammatory cytokines IL-1α, IL-1β, and their natural antagonist, the IL-1 receptor antagonist (IL-1Ra). The latter blocks IL-1 action by competing for its receptor (Dinarello 2009). The gene coding IL-1Ra maps the long arm of human chromosome 2 (band q14-21) (Steinkasserer et al. 1993; Grover et al. 2006). This gene presents a polymorphic VNTR in intron 2, with an 86 base pair repeated motif responsible for differences in the levels of expression of the receptor, which is reportedly associated with autoimmune diseases, ischemic stroke, and other pathologies (Worrall et al. 2007). The VNTR presents 5 allelic variants corresponding to 2, 3, 4, 5, and 6 copies of the 86 bp repeat (Tarlow et al. 1993; Grover et al. 2006). Three potential protein binding sites are located nearby this polymorphism, therefore the number of repeats may influence the level of gene transcription and posterior translation to a protein product.

Studies on North American U.S. population proposed the association of the allele IL-1Ra*2 of this VNTR with chronic inflammatory processes and pain (Joos et al. 2001; Foster et al. 2004), and this allele has also been associated as a risk factor for various autoimmune diseases (Rider et al. 2000; You et al. 2007; Havemose-Poulsen et al. 2007).

Since genetic diversity at the immune system is primarily important for human population's survival, the variation of this polymorphism in a group of native people inhabiting the South American Gran Chaco was explored.

A sample of 169 Amerindians from the Gran Chaco region -including Argentina and Paraguay, and a sample of 107 non-Amerindian Argentineans mainly of European ancestry (from Misiones and Buenos Aires provinces, Argentina), were analyzed. The Amerindians, identified as native according to their self designation, and their geographic location, comprised individuals belonging to five native groups from Gran Chaco: Lengua (from Paraguay, n = 36), Ayoreo (from Paraguay, n = 41), Chorote (from Santa Victoria Este, Salta province, Argentina, n = 20), Wichí (also from Santa Victoria Este, n = 39), and Mocoví (from Colonia Dolores, Santa Fe province, Argentina, n = 33). A small sample of 26 non-Amerindian Argentinians of full Japanese ancestry was also genotyped, however this group was not polymorphic for this VNTR, showing only homozygote individuals for allele 4, probably due to the small number of individuals analyzed, therefore this sample was not included in the comparative analysis.

The VNTR was amplified using the primers Fw: CTCAGCAACACTCCTAT, and Rv: TCCTGGTCTGCAGGTAA, in a MPI-Evo02 Thermocycler (La Plata, Argentina). Cycling conditions included 36 cycles of 94° 40 sec., 57° 1 min., and 72° 1 min., with an initial denaturation of 94° 2 min., and a final extension of 72° 5 min. Alleles were defined in 2% agarose electrophoresis, as in Foster et al. (2004). After post-gel DNA staining with GelGreen™ (Biotium, Hayward, CA), the electrophoretic bands were visualized in an image analyzer GelDocXR (Biorad, USA).

Allele frequencies, gene diversity, exact test of Hardy-Weinberg equilibrium, molecular variance (AMOVA), and pairwise genetic distances measured as Wright index Fst and Rst, were analyzed with Arlequin v. 3.5 (Excoffier et al. 2010). The Fst index estimates the amount of genetic differentiation between populations, by comparing total heterozygosity (five populations together) to each population heterozygosity. Genetic distances were represented using a matrix of distance MDS (multi dimensional scaling) with the program NTSYS 2.1 (Exeter) using Rst Slatkin's estimation from allele frequencies.

We found three out of the five known alleles of this VNTR (Table 1). The allele 2 was more frequent among the native people than in European and North American populations, and the allele 5 was found only in non-native European ancestry people.

Table 1. Sample size and genotype frequencies observed for IL1-Ra VNTR. Alleles are named by the number of repeats

Population	Sample size	Genotype Frequencies					
		2/2	2/4	2/5	4/4	4/5	5/5
Lengua	36	0.111	0.278	-	0.611	-	-
Ayoreo	41	0.024	0.195	-	0.780	-	-
Chorote	20	0.2	0.1	-	0.7	-	-
Wichí	39	0.205	0.154	-	0.641	-	-
Mocoví	33	0.061	0.182	-	0.757	-	-
Buenos Aires	104	0.038	0.010	-	0.875	0.019	0.058
Misiones	77	0.065	0.078	0.013	0.740	0.013	0.091
Japanese	26	-	-	-	1	-	-

Total native heterozygosity was 19%, while non-native people showed a much lower value, 2,5%. Genotypic distribution did not fit Hardy-Weinberg equilibrium among Chorote and Wichí Amerindians (Chorote: observed heterozygosity = 0.100, expected heterozygosity = 0.385, P = 0.00278 ± 0.00005; Wichí: observed heterozygosity = 0.154, expected heterozygosity = 0.410, P = 0.00019 ± 0.00001). It is noteworthy that the VNTR did not fit populational equilibrium in two cases, mostly due to the presence of higher frequencies of homozygotes for allele 2 than expected (Table 2). The genotyping of this marker was rechecked for those homozygotes, and such genotypes were confirmed.

Table 2. Allele frequencies for the five Amerindian ethnic groups (frequencies +/-S.D.)

Allele	Mocoví	Lengua	Ayoreo	Chorote	Wichí
2	0.152+/-0.044	0.250+/-0.051	0.122+/-0.036	0.250+/-0,069	0,282+/-0,051
4	0.848+/-0.044	0.750+/-0.051	0.878+/-0.036	0.750+/-0,069	0,718+/-0,051

Pairwise Fst analysis showed significant values for non-native (Buenos Aires + Misiones) compared to native populations, except for Ayoreo, and also between Wichí and Ayoreo (Table 3, data below diagonal). On the other hand, for pairwise Rst estimations non significant values were observed (Table 3, data above diagonal). Rst value between Wichí and non-native gave a limit probability of P = 0,5, showing a tendency to differentiation.

Table 3. Population pairwise Fst and Rst values below and above the diagonal, respectively. Significant results are in bold (α = 0.05)

Population pairwise Fst \ Population pairwise Rst		Non-native	Wichí	Chorote	Ayoreo	Lengua	Mocoví
	Non-native		0.01555	-0.00278	-0.00242	0.00310	-0.00898
	Wichí	**0.12312**		-0.01662	0.06532	-0.01088	0.03460
	Chorote	**0.09339**	-0.01662		0.04015	-0.01983	0.01135
	Ayoreo	0.00760	**0.06532**	0.04015		0.04112	-0.01008
	Lengua	**0.09222**	-0.01088	-0.01983	0.04112		0.01524
	Mocoví	**0.01772**	0.03460	0.01135	-0.01008	0.01524	

An analysis of molecular variance (AMOVA) (Table 4) showed a significant 3.77% differentiation between native and non-native people (P = 0.03773).

Table 4. AMOVA Analysis (distance method: Fst). Samples for testing genetic structure were grouped as native and non-native populations. α = 0.05

	Sum of squared	Variance components	Percentage of variation
Among groups	2,143	0.00565	3.77
Among populations within groups	1.443	0.00329	2.20
Within populations	76.869	0.14079	94.03
Total	80.455	0.14973	100

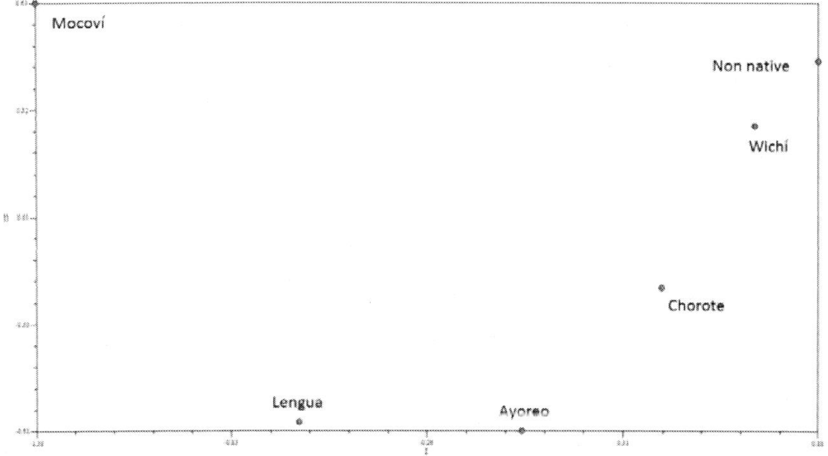

Figure 2. Multidimensional scaling (MDS) obtained from Rst distance matrix (Stress value = 0.0000). Non-native sample includes populations from Buenos Aires and Misiones.

Multidimensional map obtained from Rst distances (Figure 2) did not show any clustering, but the distribution of native populations showed certain agreement with their geographic distribution, with Mocoví (the population in a southernmost location) positioned distant from all other populations, Lengua near Ayoreo, and Chorote near Wichí. It is interesting to remark the close position of Wichí and non-native people, consistent with a migrational process. However, Fst values suggest the migrational process is more likely occurring among these native populations.

DISCUSSION

The existence of an enormous drift in American native people, acting on small subpopulations and generating variation from one population to another, has been emphasized. The stochastic accumulation of differences through the time often increases differences among populations, while intrapopulation variation decreases enormously (Cavalli-Sforza et al. 1996).

Our results on a VNTR polymorphism within a coding region might be interpreted in different ways. On the one hand, natural selection might be acting on these populations favouring proinflammatory allele 2 in order to increase a particular inflammatory-immune pathway, since many individuals belonging to native Gran Chaco populations are still living under strict environmental conditions. However, it is more likely to interpret these results as the effect of finite population size making variability to be lost rapidly.

The decrease in heterozygosity of each population compared to the diversity of the pooled native sample taken together, and the low variability observed within populations are consistent with a genetic drift process (Holsinger 2015).

When populations are isolated from one another, as it happens especially with Wichí people, they will tend to diverge from one another as a result of genetic drift. This tendency operates much faster in populations with few individuals, driving them to divergence (Hedrick 2000).

The influence of genetic drift could be detrimental to perform association genetic studies for specific diseases, because case-control and/or cohort studies might drive to erroneous conclusions when comparing populations which are substructured and in disagreement with Hardy-Weinberg equilibrium (Acosta et al. 2012).

However, the association of this allele with inflammatory disease previously demonstrated for other populations of the world, might not be acting in the same way for native people, probably due to local adaptation. This would indicate that the allele 2 probably does not influence negatively on individuals of native origin who have homozygous genotype 2-2. On the contrary, few records on inflammatory disease are available for the native people (Trujillo Miriam, personal communication). It seems that the increment on allele 2 is not related to any adaptive process but to genetic drift, that changes randomly the allele frequencies of different genetic regions along the genome. The effect of genetic drift has already been demonstrated with genetic markers located in autosomes, X and Y chromosomes (see above).

These results indicate that we must be very cautious when studying populations that passed through a process of genetic drift, which can become a confounding factor not only in evolutionary studies, but in epidemiological studies. In the first case, it could be considered as a positive selection effect; in the second one, the existence of a genetic association with certain phenotype could be falsely interpreted (Solovieva et al. 2004).

This information will contribute to a future understanding of the association of this polymorphism with disease, and its incidence in different ethnic groups.

REFERENCES

Acosta O, Solano L, Huerta D, Oré D, Sandoval J, Figueroa J, Fujita R. 2012. Variabilidad genética de la respuesta inflamatoria. I. Polimorfismo-511 C/T en el gen IL1β en diferentes subpoblaciones peruanas. *Anales de la Facultad de Medicina*, Vol. 73, Núm. 3.

Avena S., Via M., Ziv E., et al. (2012). Heterogeneity in Genetic Admixture across Different Regions of Argentina. *PloS ONE* 7(4): e34695.

Bailliet G, Rothhammer F, Carnese FR, Bravi CM, Bianchi NO. 1994. Founder mitochondrial haplotypes in Amerindian populations. *Am. J. Hum. Genet.* 54:27-33.

Bailliet G., Santos M., Alfaro E., Dipierri J., Demarchi D., Carnese F., Bianchi N. (2007). Allele and genotype frequencies of metabolic genes in Native Americans from Argentina and Paraguay. *Mutat. Res. Mar* 5;627(2):171-7.

Bourgeois S, Yotova V, Wang S, et al. 2009. X-Chromosome Lineages and the Settlement of the Americas. *Am. J. Phys. Anthropol.* 140: 417–428.

Braunstein, J. 2006. El signo del agua. Formas de clasificación étnica Wichí. Definiciones étnicas, organización social y estrategias políticas en el Chaco y la Chiquitania. Editor: Isabelle Combes. Actes et Memoire de l'Institute Français d'Etudes Andines 11.

Catanesi C, Martina P, Giovambattista G, Zukas P, Vidal Rioja L. 2007. Geographic structure in Gran Chaco Amerindians based on five X-STRs. *Hum. Biol.* 79 (4): 463-474.

Catanesi C.I., Méndez M.G., Salceda S.A., Vidal Rioja L.A. 2001. Perfil morfométrico-molecular de tres poblaciones nativas sudamericanas. Milenio. M. Amanda Caggiano Editor, Chivilcoy (Argentina) 127-130. ISBN 987-98845. Edición no periódica con referato.

Catanesi C.I., Tourret M.N., Vidal Rioja L.B. 2001. Genetic diversity at LPL and CD4 microsatellite loci in South American populations. Journal of Basic and Applied Genetics XIV, 1: 41-46. Editada por la Sociedad Argentina de Genética (SAG). Argentina. ISSN BAG 1666-0390.

Cavalli-Sforza, L., P. Menozzi, A. Piazza. 1996. In: The History and Geography of Human Genes. Princeton NJ: Princeton University Press.

Citro S. 2006. La Fiesta del 30 de Agosto entre los mocovíes de Santa Fe. Buenos Aires, Facultad de Filosofía y Letras, Universidad de Buenos Aires.

Dejean CB, Crouau-Roy B, Goicoechea AS, Avena SA, and Carnese FR. 2004. Genetic variability in Amerindian populations of Northern Argentina. *Genetics and Molecular Biology*, 27, 4, 489-495.

Demarchi DA, Mitchell RJ. 2004. Genetic Structure and Gene Flow in Gran Chaco Populations of Argentina: Evidence from Y-Chromosome Markers. *Human Biology*, v. 76, no. 3, pp. 413–429.

Demarchi DA, Panzetta-Dutari GA, Motran CC, Marcellino AJ. 2001. Mitochondrial DNA haplogroups in Amerindian populations from the Gran Chaco. *Am. J. Phys. Anthropol.*, Vol. 115, N. 3: 199-203.

Dinarello CA: Immunological and inflammatory functions of the interleukin-1 family. *Annu. Rev. Immunol.* 2009, 27:519-550.

Dinarello CA, Simon A, van der Meer JW. Treating inflammation by blocking interleukin-1 in a broad spectrum of diseases. *Nature Rev Drug Discov.* 2012; 11:633–652.

Excoffier L Lischer H. 2010. Arlequin suite ver 3.5: a new series of programs to perform population genetics analyses under Linux and Windows. *Mol. Ecol. Resources*, Vol. 10, Issue 3: 564–567.

Franceschi ZA and Dasso MC. 2010. Etno-Grafías, La escritura como testimonio entre los Wichí. El universo Wichí: historia y cultura. Ed. Corregidor, Buenos Aires 336 pp.

Foster D.C., Sazenski T.M., Stodgell C.J. 2004. Impact of genetic variation in interleukin-1 receptor antagonist and melanocortin-1 receptor genes on vulvar vestibulitis syndrome. *The Journal of Reproductive Medicine*, 49 (7): 503-509.

Gabay C, Lamacchia C, Palmer G. IL-1 pathways in inflammation and human diseases. *Nature Rev. Rheumatol.* 2010; 6:232–241.

García A, Demarchi DA. 2009. Incidence and distribution of Native American mtDNA haplogroups in central Argentina. *Hum. Biol.* 81(1):59-69.

Garlanda C, Dinarello CA, and Mantovani A. 2013. The interleukin-1 family: back to the future. *Immunity.* 2013 December 12; 39(6): 1003–1018.

Glesmann L.A., Martina P.F., Méndez M.G., Catanesi C.I. 2011. High genetic variation in Y chromosome patterns of the Mocoví population. *Rev. del Museo de Antropología,* vol. 4: 179-186.

Glesmann L.A., Martina P.F. Catanesi C.I. 2013. Genetic variation of X-STRs in the Wichí population from Chaco province, Argentina. Human Biology. Wayne State Univ. Press, Detroit. Paper 32.

Goicoechea AS, Carnese FR, Dejean CB, Avena SA, Weimer TA, Franco MH, Callegari-Jacques S, Estalote A, Simoes ML, Palatnik M, Salomoni P and Salzano FM. 2001. Genetic relationships between Amerindian populations of Argentina. *Am. J. Phys. Anthropol.* 115:133-143.

Grover S, Tandon S, Misra R, Aggarwal A. 2006. Interleukin-1 receptor antagonist gene polymorphism in patients with rheumatoid arthritis in India. *Indian J. Med. Res.* 123(6): 815-820.

Havemose-Poulsen A., Sørensen L.K., Bendtzen K., Holmstrup P. 2007. Polymorphisms within the IL-1 gene cluster: effects on cytokine profiles in peripheral blood and whole blood cell cultures of patients with aggressive periodontitits, juvenile idiopathic arthritis, and rheumatoid arthritis. *J. Periodontol.* 78: 475-492.

Hedrick PW. 2000. Genetics of population, 2nd. Ed. Jones and Bartlett Publishers. USA, 553 pp.

Holsinger KE. 2015. Mutation, migration and genetic drift. In: Lecture notes in population genetics. Department of Ecology & Evolutionary Biology, University of Connecticut 333 pages.

Horai S, Kondo R, Nakagawa-Hattori Y. 1993. Peopling of the Americas, founded by four major lineages of mitochondrial DNA. *Mol. Biol. Evol.* 10:23-47.

Joos L., McIntyre L., Ruan J., Connett J.E., Anthonisen N.R., Weir T.D., Paré P.D., Sandford A.J. 2001. Association of IL-1β and IL-1 receptor antagonist haplotypes with rate of decline in lung function in smokers. *Thorax,* 56: 863-866.

Jurado Medina LS, Ramallo V, Calandra H, Lamenza G, Braunstein J, Salceda S, Bailliet G. 2014. Gran Chaco paternal lineages, a DNA approach. Folia Historica del Nordeste N° 22, Resistencia, Chaco, IIGHI - III-CONICET/UNNE: 187-202.

Martínez Sarasola C. 2004. Las comunidades que ocupaban nuestro territorio en el siglo XVI. Nuestros Paisanos los Indios. Vida, historia y destino de las comunidades indígenas en la Argentina. Editorial EMECE, Buenos Aires, Argentina. I:77-82.

Motti JMB, Muzzio M, Ramallo V, Rodenak Kladniew B, Alfaro EL, Dipierri JE, Bailliet G, Bravi CM. 2013. Origen y distribución espacial de linajes maternos nativos en el noroeste y centro oeste argentinos. *Rev. Arg. de Antropología Biológica*, Vol. 15, N. 1: 03-14.

Mulligan C., Hunley K., Cole S., Long J. 2004. Population genetics, history, and health patterns in native Americans. *Annual Review of Genomics and Human Genetics* 5: 295-315.

Pauro M, García A, Bravi CM, Demarchi DA. 2010. Distribución de haplogrupos mitocondriales alóctonos en poblaciones rurales de Córdoba y San Luis. Rev. *Argentina de Antropología Biológica* Vol. 12, N. 1: 47-55.

Ramallo V, Santos MR, Muzzio M, Motti JMB, Salceda S, Bailliet G 2009. Linajes masculinos y su diversidad en comunidades Wichí de Formosa. Revista del Museo de Antropología, Facultad de Filosofía y Humanidades – Universidad Nacional de Córdoba – Argentina, 2: 67-74.

Ribeiro-Rodrigues E., de Jesus Brabo Ferreira Palha T., Auler Bittencourt E., Ribeiro-dos-Santos A,. Santos S. 2011. Extensive survey of 12 X-STRs reveals genetic heterogeneity among Brazilian populations. *Int. J. Legal Med.* 125:445–452.

Ribeiro Rodrigues E., Dos Santos N., Campos Ribeiro dos Santos A., et al. 2009. Assessing Interethnic Admixture Using an X-Linked Insertion-Deletion Multiplex. *Am. J. Hum. Biol.* 21:707–709.

Rider L.G., Artlett C.M., Foster C.B., Ahmed A., Neeman T., Chanock S.J., Jimenez S.A., and Miller F.W. 2000. Polymorphisms in the IL-1 receptor antagonist gene VNTR are possible risk factors for juvenile idiopathic inflammatory myopathies. *Clin. Exp. Immunol.* Jul 2000; 121(1): 47–52.

Rondón F., Oosorio J., Peña A., Garcés H., Barreto G. 2008. Diversidad genética en poblaciones humanas de dos regiones colombianas. *Colomb Med.* 39 (Supl 2): 52-60.

Salzano F. and Bortolini M. 2002. The Evolution and Genetics of Latin American Populations. Cambridge University Press, Cambridge, 512 pp.

Schurr TG, Ballinger SW, Gan YY, Hodge JA, Merriwether DA, Lawrence DN, Knowler WC, Weiss KM, Wallace DC. 1990. Amerindian mitochondrial DNAs have rare Asian mutations at high frequencies, suggesting they derived from four primary maternal lineages. *Am. J. Hum. Genet* 46(3):613-623.

Sims JE, Smith DE. 2010. The IL-1 family: regulators of immunity. *Nat. Rev. Immunol.* 2010; 10:89–102.

Solovieva S, Leino-Arjas P, Saarela J, Luoma K, Raininko R, Riihimaki H. 2004. Possible association of interleukin 1 gene locus polymorphisms with low back pain. *Pain*, 2004 May;109(1-2):8-19.

Steinkasserer A., Spurr N.K., Sim R.B. 1993. Chromosomal mapping of the human interleukin-1 receptor antagonist gene (IL-1RN) and isolation of specific YAC clones. *Inflammation Research* 38(2): C59-C60.

Tarlow JK, Blakemore AI, Lennard A, Solari R, Hughes HN, Steinkasserer A, Duff GW. Polymorphism in human IL-1 receptor antagonist gene intron 2 is caused by variable numbers of an 86-bp tandem repeat. *Hum. Genet.* 1993 May;91(4):403-4.

Tissera R. d. l. M. 2008. Chaco, Historia General. Subsecretaría de Cultura, Min. De Educación, Cultura, Ciencia y Tecnología de la provincia de Chaco. Librería de la Paz, Resistencia, Chaco, Argentina, 46, 401, 404.

Torroni A, Schurr TG, Cabell MF, Brown MD, Neel JV, Larsen M, Smith DG, Vullo CM, Wallace DC. 1993. Asian affinities and continental radiation of the four founding Native American mtDNAs. *Am. J. Hum. Genet.* 53:563-590.

Tourret M.N., Catanesi C.I., Vidal Rioja L.B. 2000. Variability of the F13B locus in South American Populations. Human Biology, v.72, 4, 707-714. Editorial: Wayne State University Press, Detroit, Michigan, USA. ISSN 0018-7143.

Wang S, Ray N, Rojas W, Parra MV, Bedoya G, Gallo C, Poletti G, Mazzotti G, Hill K, Hurtado AM, Camrena B, Nicolini H, Klitz W, Barrantes R, Molina JA, Freimer NB, Bortolini MC, Salzano FM, Petz-Erler ML, Tsuneto LT, Dipierri JE, Alfaro EL, Bailliet G, Bianchi NO, Llop E, Rothhammer F, Excoffier L, Ruiz-Linares A. 2008. Geographic patterns of genome admixture in Latin American mestizos. *PLoS Genet.* 4(3):e1000037.

Worrall B.B., Brott T.G., Brown R.D., Brown W.M., Rich S.S., Arepalli S., Wavrant-De Vrièze F., Duckworth J., Singleton A.B., Hardy J., Meschia J.F., et al. 2007. IL1RN VNTR polymorphism in ischemic stroke: analysis in 3 populations. *Stroke.* 2007 Apr;38(4):1189-96.

Yang NN, Mazieres S, Bravi C, Ray N, Wang S, Burley MW, Bedoya G, Rojas W, Parra MV, Molina JA, Gallo C, Poletti G, Hill K, Hurtado AM, et al. 2010. Contrasting Patterns of Nuclear and mtDNA Diversity in Native American Populations. *Ann. Hum. Genet.* 2010 Nov;74(6):525-38.

You C., Li J., Xie X., Zhu Y., Li P., Chen Y. 2007. Association of interleukin-1 genetic polymorphisms with the risk of rheumatoid arthritis in Chinese population. *Clin. Chem. Lab. Med.* 45: 968-971.

Zago, M.A., W.A. Silva. Jr., M.H. Tavella, S.E.B. Santos, J.F. Guerreiro and M.S. Figueiredo. 1996. Interpopulational and intrapopulational genetic diversity of Amerindians as revealed by six variable number of tandem repeats. *Hum. Hered.* 46:274-289.

INDEX

#

20th century, 75, 79, 92

A

access, 58, 65
accommodation, 34, 67
accounting, 77
acid, 80, 81
acquaintance, 5
action potential, 81
adaptability, vii, viii, 61, 62, 63, 67, 69, 71, 73, 75, 76, 78, 92
adaptation(s), ix, 2, 5, 14, 15, 47, 67, 102, 112
adenine, 81
adjustment, 29
adulthood, 8
adults, viii, 62, 68
advancements, 47
age, 16, 17, 19, 20, 24, 30, 51, 68, 69, 70, 71, 72, 78
age of reptiles, 37
aggression, 68, 80, 94
agriculture, 66
alanine, 81
allele, ix, 72, 73, 80, 102, 107, 108, 109, 110, 112
Amerindians, ix, 101, 108, 110, 113, 118

amino, 80, 81
ancestors, 26, 43, 44
antibiotic, 15
antigen, 80
apoptosis, 68
architect, 28
Argentina, ix, 101, 103, 106, 108, 109, 113, 114, 115, 116, 117
arginine, 80
Aristotle, 53
arithmetic, 21, 45
arrow of time, 25
arteries, 65
arthritis, 115
assessment, 4, 9, 19, 20, 24, 27, 29, 53, 54
autoimmune disease(s), 68, 80, 108
awareness, 51

B

back pain, 117
bacteria, 15
bacterial colonies, 15
base, 29, 35, 37, 66, 108
base pair, 108
benefits, 6
biological consequences, 65
biosphere, 59
birth rate, 64
blood, 65, 72, 97, 104, 115

BMI, 70
body size, 6, 22, 39
Bolivia, 103
bone, 17, 21
brain, 6, 18, 50
branching, 41
breeding, 18, 66

C

caloric intake, 66
cancer, 65
carbohydrates, 66
carbon, 52
carbon dioxide, 52
cardiovascular disease(s), 64, 67
cattle, 66
causality, 18
cell culture, 115
chaos, 25
Chicago, 59, 60
children, 50, 51, 52, 64, 66, 74, 75
chimpanzee, 54
chromosome, 80, 81, 104, 105, 108
CIS, 96
cities, 45
citizens, 64, 69, 78, 80, 82, 83, 84, 85
classification, 68
climate, 2, 34, 49
clustering, 111
coding, 67, 104, 107, 108, 112
codon, 81
combined effect, 94
communit(ies), 4, 45, 64, 107
comparative analysis, 81, 84, 109
compensation, 93
competition, viii, 2, 21, 39, 40, 41, 44, 48, 65
complement, 5
complex interactions, 104
complexity, vii, 1, 2, 4, 5, 7, 19, 25, 26, 27, 28, 29, 30, 31, 32, 33, 34, 35, 36, 37, 38, 39, 40, 41, 42, 43, 44, 45, 46, 47, 48, 49, 53, 54, 55, 56, 57, 58, 59
complications, 97

comprehension, 25
conception, 4, 8, 30, 44, 91
concordance, 48, 52
condensation, 7, 8, 9, 10, 13, 14, 19, 20, 21, 22, 23, 24, 25, 50, 51, 52, 56
congress, 96, 99
conservation, 58
construction, viii, 2, 18, 28, 31, 46, 56, 65
consumption, 66
control group, 69
controversial, 28, 46, 49, 53
conviction, 16
correlation, 107
cortical neurons, 23
cough, 65
cranium, 6
critical period, 13
crocodilians, 36
cultural evolution, 1, 2, 5, 21, 35, 39, 43, 44, 45, 46, 47, 48, 53, 54, 58
culture, vii, 1, 5, 43, 44, 45, 46, 47, 48, 49, 54, 55
curricula, 53
cycles, 72, 109
cytokine IL-1, ix, 101
cytokines, 108

D

Darwin, Charles, 4, 40
Darwinian evolution, 49, 51
Darwinism, 43, 54
deaths, 63
defects, 64, 67
deficiency, viii, 4, 62, 84, 91
demography, 95
denaturation, 72, 109
denial, 4
Department of Education, 1
deposits, 66
depression, 64
depth, 25
destruction, 68
detectable, 36
developed countries, 63, 64, 65

developmental change, 5, 58, 59
developmental process, viii, 1, 2, 3, 5, 7, 15, 16, 23, 24, 28, 52, 56
diabetes, vii, viii, 61, 62, 68, 75, 96, 97, 98, 99, 100
diamonds, 30, 32
diet, 67
digestion, 10
dignity, 55, 58
direct observation, 17
directionality, 6
disaster, 13
disease gene, 79
diseases, ix, 2, 63, 64, 65, 67, 69, 73, 91, 95, 96, 97, 101, 107, 112, 114
disequilibrium, 106
disorder, 80, 94
displacement, 7, 19, 20
distribution, ix, 63, 66, 67, 73, 74, 80, 81, 83, 84, 85, 86, 88, 89, 93, 94, 101, 105, 106, 109, 111, 115
divergence, 18, 58, 112
diversity, vii, 3, 7, 41, 42, 44, 63, 96, 102, 104, 105, 107, 109, 112, 114
DNA, 3, 18, 26, 28, 72, 104, 109, 114, 115
doctors, 97
domestication, 45
drawing, 17

entropy, 25, 26
environment(s), 2, 7, 12, 13, 14, 15, 22, 23, 24, 33, 44, 47, 49, 50, 56, 63, 67, 96, 97
environmental conditions, 46, 67, 106, 107, 112
environmental factors, 5, 65, 68
epidemiology, 97
epistemology, 52
equilibrium, 109, 110, 112
ethnic groups, ix, 102, 104, 105, 106, 110, 113
evidence, 73, 79
evolution, vii, viii, 1, 2, 3, 4, 5, 6, 7, 8, 13, 15, 16, 17, 18, 21, 22, 23, 24, 25, 26, 27, 28, 29, 30, 31, 32, 33, 34, 35, 36, 37, 38, 39, 40, 41, 42, 43, 44, 45, 46, 47, 48, 49, 50, 51, 52, 53, 54, 55, 56, 57, 58, 59, 61, 62, 63, 65, 66, 67, 91, 94, 96
evolutionary human genetics, vii, 63
examinations, 24, 69
exercise, 51
exons, 80
exponential functions, 29, 31, 33
external environment, vii, 1, 5, 13, 14, 24, 33
extinction, 13, 14, 15, 33, 42
extraction, 72

E

earthworms, 54
ecology, 33
economic progress, 63
editors, 25
education, 50, 51, 52, 106
election, 64
electrophoresis, 109
embryogenesis, 6, 15, 24, 73
embryology, 24
encoding, 108
endocrine, 69, 71
endocrinology, 98
energy, 25, 66
energy consumption, 66

F

family planning, 73
farmers, 64
fat, 66
fertility, 62, 63, 64, 71
Finland, 83
fishing, 103
fluctuations, 62, 64
food, 2, 42, 66, 68, 91
force, 25, 48, 62, 105
formation, viii, 6, 22, 28, 29, 33, 36, 39, 51, 62, 92, 95
formula, 10, 13, 14, 20, 31, 71
fossils, 36
fragments, 72, 73
fungus, 26

G

gel, 72, 109
gene(s), viii, ix, 6, 8, 28, 31, 35, 44, 47, 48, 49, 55, 57, 60, 61, 62, 63, 66, 68, 69, 70, 72, 76, 79, 80, 81, 82, 83, 84, 85, 86, 87, 88, 89, 90, 91, 92, 93, 94, 99, 101, 104, 107, 108, 109, 113, 114, 115, 116, 117
gene pool, 31, 63
genetic background, 5
genetic code, 2
genetic diversity, 104, 108, 118
genetic drift, vii, ix, 101, 102, 105, 106, 107, 112, 113, 115
genetic factors, 80, 92
genetic information, 33
genetic marker, ix, 102, 113
genetics, vii, 25, 36, 63, 67, 79, 80, 95, 97, 114, 115, 116
genome, ix, 28, 48, 67, 102, 113, 117
genotype, ix, 62, 66, 101, 109, 112, 113
genotyping, 110
geometry, 45
gestation, 24
gland, 65
glucose, 97
graph, 31, 44
greenhouse, 52
grouping, 106
growth, 3, 8, 10, 12, 15, 22, 23, 28, 33, 35, 39, 42, 44, 45, 48, 50, 56, 58, 63, 68, 76, 78, 79, 91, 92, 96
guanine, 81

H

habitat, 50
haplotypes, 105, 113, 115
hazards, 49
health, 68, 69, 75, 79, 92, 93, 96, 97, 99, 116
health care, 68, 69, 75, 79, 92, 93
heart disease, 66
hegemony, 46, 51
height, 39
heredity, 44, 94
heritability, 92
heterogeneity, viii, 61, 69, 79, 92, 95, 113, 116
history, 3, 4, 13, 16, 18, 21, 22, 25, 27, 31, 46, 52, 56, 64, 103, 116
HLA, 80, 83, 93, 94, 104, 107
homeostasis, 83
hominids, 42
homozygote, 83, 85, 86, 94, 107, 108
House, 60
human, vii, viii, 1, 2, 5, 6, 16, 17, 18, 20, 21, 26, 32, 33, 39, 42, 43, 44, 45, 46, 47, 48, 49, 50, 53, 54, 55, 56, 57, 61, 63, 65, 66, 102, 108, 114, 117
human body, 46
human brain, 33, 48, 50
human development, 66
human dignity, 54
hunting, 66, 103
hypothesis, vii, 29, 63, 68

I

identification, 4
idiopathic, 115, 116
image, 109
imitation, 49
immigration, 65
immune disorders, 83
immune response, viii, 62, 80, 94
immune system, 64, 108
immunity, 117
impregnation, 73
improvements, 34, 35
incidence, ix, 102, 113
independence, 13, 69, 89, 91, 94
India, 115
individual development, 2, 5, 23, 37, 56, 62
individuals, ix, 10, 33, 34, 62, 64, 66, 81, 101, 104, 105, 107, 108, 112
indoctrination, 51
inflammation, 114
inflammatory disease, ix, 102, 112

inflammatory processes, ix, 101, 107, 108
information technology, 32
inheritance, 49, 80, 92, 93, 94, 96, 105
inhibition, 65
insertion, 7
insulin, viii, 61, 66, 68, 70, 75, 78, 79, 81, 84, 91, 92, 93, 94, 97
insulin deficiency, viii, 62, 84
insulin insufficiency, viii, 62, 68, 81, 94
insulin therapy, viii, 61, 68, 70, 75, 76, 79, 91, 92, 93
intelligence, 44, 49, 53
intentionality, 8
interleukin 1 receptor, vii, 101
intron(s), ix, 80, 101, 108, 117
invertebrates, 6
ions, 81
isolation, 106, 117
issues, 63

J

justification, 28, 31, 43

K

kidneys, 18, 24

L

LADA (latent autoimmune diabetes in adults), viii, 62, 68, 69, 70, 76, 79, 80, 81, 82, 83, 84, 85, 86, 88, 89, 91, 92, 93, 94
languages, 104
larval development, 6
latent autoimmune diabetes of adults, 62
Latin America, 102, 116, 117
learning, 47, 48, 50, 51
legs, 6
lifetime, 10
ligand, 81
light, 56
loci, 114
locus, 117
longitudinal study, 24
lung function, 115
lying, 39
lymphocytes, 80
lymphoid, 80

M

macrophages, 107
magnitude, 54
majority, vii, 2, 9, 55, 64, 66
mammals, 24, 39
mapping, 117
marriage, 96
maternal lineage, 117
mathematics, 52
matrix, 109, 111
matter, 27, 46
measurement(s), 25, 29, 31, 42, 56
medical, 64, 65, 96, 98, 107
medicine, viii, 61, 63, 65, 67, 92, 96
mellitus, vii, viii, 61, 62, 97, 100
mental development, 52
Metabolic, 97
metabolic syndrome, 97
metabolism, 98
metaphor, 18, 28, 37, 39, 41
methodology, 63
microorganisms, 65
microsatellites, 104
migration, 62, 96, 115
mitochondrial DNA, 104, 106, 115, 116
mixing, 66
models, 36, 94
modern society, 65, 66
modifications, 3, 6, 18, 40
molecular biology, 3
morbidity, 71, 91
mortality, 63, 64, 75
Moscow, 64, 72, 95, 96, 97, 98, 100
motif, 108
MPI, 109
mtDNA, 115, 118
muscles, 66

muscular tissue, 66
mutant, 72, 73, 83, 84, 94
mutation(s), 6, 40, 62, 67, 80, 81, 85, 94,
    104, 116
mutation rate, 104
myocardial infarction, 66

N

Native Americans, 102, 106, 113
native population, ix, 101, 102, 105, 107,
    110, 111
natural science(s), 47
natural selection, vii, 1, 2, 3, 4, 5, 7, 8, 9,
    10, 13, 14, 17, 18, 21, 22, 23, 24, 26, 27,
    31, 33, 34, 35, 36, 37, 39, 41, 42, 47, 49,
    52, 56, 57, 58, 61, 62, 63, 66, 67, 73, 95,
    97, 107, 112
negative consequences, 92
neglect, 40
nervous system, 45
neurons, 48
neutral, vii, 63, 104
neutral evolution, vii, 63
next generation, 31, 33, 50
North America, 108, 109
Norway, 83
nucleotides, 104
numerical analysis, 28, 31
nutrient(s), 15
nutrition, 66

O

obesity, 67, 91
opportunities, 34, 50
organic evolution, 2, 5, 27, 34, 42, 44, 46,
    50, 56
organism, 18, 41, 62, 65
organs, 7, 18, 22, 24, 108
overlap, 83

P

Pacific, 66
pain, 65, 108
Paraguay, ix, 101, 103, 106, 107, 108, 113
parallel, 52
parallelism, 52
parasites, 40
parents, 33
pathogenesis, 93
pathology, 69, 71, 98
pathway(s), 112, 114
PCR, 72
pelvis, 6
peptide, 81
perinatal, 75
peripheral blood, 115
personal communication, 112
phenotype(s), 62, 63, 78, 113
Philadelphia, 57
physical environment, 49
physics, 52
plants, 37, 45, 46
playing, 108
Poland, 83
polymerase, 72
polymerase chain reaction, 72
polymorphism(s), ix, 67, 69, 70, 72, 80, 81,
    82, 83, 84, 85, 86, 87, 88, 89, 90, 91, 93,
    94, 101, 102, 104, 108, 112, 113, 115,
    117, 118
population, vii, viii, 1, 2, 9, 10, 12, 13, 14,
    15, 21, 23, 36, 50, 51, 61, 62, 64, 65, 66,
    67, 68, 69, 70, 71, 73, 74, 75, 76, 77, 78,
    79, 80, 84, 91, 92, 93, 95, 98, 99, 104,
    105, 106, 107, 108, 109, 111, 112, 114,
    115, 118
population density, 64
population growth, vii, 1, 9, 12
population size, 10, 12, 13, 14, 112
population structure, 77
positive feedback, 33, 57
potassium, 81
predators, 2, 10, 34
pregnancy, 73

present value, 32
preservation, 54, 64
prevalence in population, 62, 77, 78, 79, 92
principles, 43, 44
probability, 78, 79, 92, 93, 110
proband, 70, 79, 92, 93, 94
prophylactic, 100
protection, 66
protective mechanisms, 65
psychological processes, 52
public health, viii, 62, 64

Q

quantitative estimation, 63, 67
questioning, 69

R

race, 26, 34
racial differences, 46
radiation, 117
rate of change, 49
reading, 51, 95
reasoning, 18, 23, 28, 41, 42, 46
recall, 50
receptor, vii, 80, 108, 114, 115, 116, 117
recognition, 65
recombination, 105
relative adaptability, 62, 69, 71, 73, 75, 78, 92
relatives, 54
religion, 48
reproduction, 8, 9, 10, 12, 13, 14, 62, 63, 67
reproductive age, 62, 63, 64
reptile, 18
reputation, 16
requirement, 31
researchers, 2, 25
reserves, 66
resistance, 15, 65
resources, 66, 68, 91
response, 94, 108
restriction enzyme, 72

retardation, 64
rheumatoid arthritis, 115, 118
risk, 73, 97, 108, 116, 118
risk factors, 116
routes, 31, 43
Russia, 95, 96

S

scaling, 109, 111
schizophrenia, 67
school, 51, 52, 53, 98
science, vii, 1, 3, 20, 25, 45, 47, 51, 52, 53, 57, 60, 96
scientific progress, 47, 53
secretion, 81
segregation, 66
self-reinforcing feedback, viii, 1, 2, 33, 34, 35, 38, 39, 42, 44, 46, 50, 53, 56
senses, 10
services, viii, 62, 64
shape, 3, 29, 30, 59
shelter, 34, 37
showing, 17, 105, 108, 110
signs, 17, 44
SNP, 104, 105
social development, 63
social environment, 28, 49, 50
social integration, 106
social life, 2
social organization, 63
society, 45, 50, 51, 63, 65, 95, 96, 106
solution, 4, 42
South America, vii, ix, 101, 102, 107, 108, 114, 117
Spain, 70
species, vii, viii, 2, 7, 8, 9, 13, 14, 16, 17, 21, 22, 26, 34, 36, 37, 38, 39, 40, 41, 42, 43, 44, 45, 46, 47, 49, 53, 54, 55, 56, 63
spontaneous abortion, 69, 73
starvation, 66
stasis, 36
state(s), 5, 6, 14, 42, 54, 65
statistics, 71
sterile, 63

stretching, 12, 29
stroke, 108, 117
strong force, 105
STRs, 104, 105, 107, 113, 115, 116
structure, 6, 8, 23, 24, 39, 44, 59, 62, 71, 79, 92, 96, 102, 104, 111, 113
substitutions, 67, 81
substrate, 33, 44
supervision, 28
survival, 10, 12, 13, 14, 48, 49, 71, 74, 75, 76, 79, 91, 92, 108
survival rate, 71, 74, 75, 76, 79, 91, 92
survival value, 49
survivors, 36
susceptibility, viii, 62, 63, 67, 68, 69, 76, 79, 80, 81, 84, 91, 92, 93, 94
Sweden, 1, 83
symptoms, 69
syndrome, 114
synthesis, 59, 81

T

tandem repeats, 104, 118
technology, 33, 45, 46, 51
teeth, 65
temperature, 24, 72
tempo, 6, 8
tension, viii, 61, 63, 67, 68, 69, 75, 79, 91, 92, 93
territory, 102, 104, 106
testing, 92, 111
textbooks, 24
therapy, viii, 15, 61, 67, 68, 70, 75, 79, 91, 92, 93
thoughts, 37
threonine, 81
tissue, 66
traits, 6, 7, 8, 9, 10, 13, 14, 16, 17, 18, 20, 21, 22, 23, 24, 25, 34, 35, 38, 39, 46, 50, 56, 62, 65, 92
transcription, 108
transformation(s), 3, 96
translation, 52, 108
transportation, 81

treatment, viii, 15, 61, 63, 67, 97, 100, 107
tree of life, viii, 2, 36, 38, 40, 41, 42, 45, 46, 47, 56
tryptophan, 80
Tyler, John, 48
type 1 diabetes, viii, 61
type 1 diabetes mellitus, viii, 61
type 1 DM, viii, 61, 68, 69, 70, 71, 73, 74, 75, 76, 78, 79, 80, 82, 83, 84, 85, 86, 88, 89, 91, 92, 93, 94
type 2 diabetes, 97
type 2 DM, viii, 61, 68, 69, 70, 71, 74, 75, 78, 79, 80, 81, 82, 83, 84, 85, 86, 88, 89, 91, 93, 94
tyrosine, 80

U

Ukraine, 61, 83, 97, 98
underlying mechanisms, 47
uniform, 8
universities, 51
urban, 63
urban population, 63
urbanization, 64, 66, 96, 97
USA, 64, 83, 109, 115, 117

V

variables, 10, 12, 73
variations, 4, 17
vegetation, 37, 106
vein, 27
vertebrates, 6, 24, 36, 39
vocabulary, 50

W

weapons, 45
worldwide, 68

X

X chromosome, 104, 105, 106, 107

Y

Y chromosome, ix, 102, 104, 106, 113, 115
YAC, 117
young people, 50